EFFECTIVE THEORY OF QUANTUM GRAVITY: SOLUBLE SECTOR

Subodha Mishra
Institute of Technical Education & Research
Bhubaneswar, India

Joy Christian
Wolfson College, University of Oxford
Oxford, United Kingdom

Published 2011 by abramis

www.abramis.com

ISBN 978 1 84549 476 6

© Subodha Mishra & Joy Christian 2011

Printed and bound in the United Kingdom

Abramis is an imprint of arima publishing.

arima publishing
ASK House, Northgate Avenue
Bury St Edmunds, Suffolk IP32 6BB
t: (+44) 01284 700321

www.arimapublishing.com

Contents

III   Relativization of the problem :Effective Theory of Quantum Gravity   115

PREFACE

It is well known that Einstein's General Theory of Relativity (GTR)cannot be covariantly quantized as a field theory, but it is now known that the non-special relativistic limit of GTR, the so-called Newton-Cartan theory of gravity, can bequantized exactly. In fact the resulting Hamiltonian represents an exactly soluble sector of the full quantum gravity.

This book establishes an effective theory of quantum gravity stemming from this fact. It is a theoretical study of exact quantization of the gravitational interaction for Newton-Cartan-Schrödinger theory, its special relativization, and its application to the self-gravitating systems like neutron stars, white dwarfs, black holes and also the universe. The special relativization of the theory is accomplished using Mach's principle in case of universe, and using Schwarzschild radius in the case of other self-gravitating systems like stars. Through a variational approach using a trial local density for the self-gravitating systems we further show how one can calculate the ground state energy and derive other parameters like size, density, Chandrasekhar limit (in the case of white dwarfs), and maximum mass of a neutron star. In the case of universe we obtain the Eddington number (number of baryons in the visible universe), mass density, time variation of Gravitational constant, and an upper limit on the mass of neutrinos. The ratio of baryons to photons and the number of photon per unit volume are also predicted reasonably well within our approach. We also derive a relation connecting the cosmological constant, time and the temperature of the cosmic background radiation of our acceleratedly expanding universe. When the value of cosmological constant is assumed to be vanishing, we recover Gamow's relation with a much better coefficient. Otherwise our relation predicts a reasonably correct value for the cosmological constant.

ACKNOWLEDGMENTS

SM acknowledges the supports he got as grants from scholarships awarded to him in dierent phases of his carrier when various aspects of the work of this book were carried out. These include the doctoral fellowship at Institute of Physics, India; G. Ellseworth Huggins Scholarship by the Graduate School, University of Missouri-Columbia, USA and O.M.Stewart Scholarships by Department of Physics and Astronomy of the same university. He remembers his collaboration with D. N. Tripathy who is no more. He appreciates with thanks the valuable encouragement he received from Henry White, Bahram Mashhoon and H. R. Chandrasekhar in USA and from Pitabas Pradhan, P. C. Naik, S. M. Bhattacharjee and V. S. Ramamurthy in India.

JC is indebted to Jeremy Buttereld, Abner Shimony, Jeeva Anandan, Paul Busch, Ashwin Srinivasan, Andreas Fuentes, Wolfram Latsch, Rui Camacho, Savitaben Christian, and The Mrs L.D. Rope Third Charitable Settlement for their generous and timely support without which the original work (published in the Physical Review D) would not have materialized. He is also grateful for the generous interest shown by Roger Penrose in his precursory ideas on the subject.

We are also grateful to Alwyn van der Merwe, ex-chief editor of Foundations of Physics for his keen interest in our work and support to get our work published as a book. We thank R. Franklin of Abramis, UK for publishing our book.

Part I

Quantum Gravity

Chapter 1

Effective Theory of Quantum Gravity

1.1 Introduction

The elusive theory of quantum gravity [1, 2] is the theory that can describe the quantized nature of gravitational interaction. In this work we establish an effective theory of quantum gravity. The effective theory is derived [3] after we special relativize the quantized Newton-Cartan-Schrodinger theory [4]. The N-C theory is the true Newtonian limit $(c \rightarrow \infty)$ of the General Theory of Relativity. It is known that General Theory of Relativity (GTR) can not be covariantly quantized due to technical problem of dealing with 'time'. But (Fig. 1) we take a detour to reach at the Quantum Theory of Gravity. Since the relativization is done after we quantize the N-C-S theory, we call it an effective theory. Also, since gravity is the dominant interaction at large scales, quantum cosmology [5] should be based on the theory of quantum gravity. The effective theory of quantum gravity not only takes into account the consistent quantization of gravity but also explains cosmology and other self-gravitating objects successfully. It is interesting to note that after we quantize the connection of spacetime in the N-C theory with vanishing spatial curvature, we get a Hamiltonian which is purely quantum mechanical with a fixed back-ground space-time manifold. The

Hamiltonian turns out to be that of a self-gravitating system.

Clearly, unlike the lopsided orthodox approaches towards a putative quantum theory of gravity, this is a fairly 'evenhanded' approach. For in the orthodox approaches, quantum superpositions [6] are indeed presumed to be sacrosanct at all physical scales, but only at a very high price of some radical compromises with Einstein's theory of gravity e.g., at price of having to fix both the topological and differential structures of spacetime *apriory*, as in the 'loop quantum gravity' programme, or -even worse- at a price of having to assume some $non - dynamical$ causal structure as a fixed arena for dynamical processes, as in the current voguish 'M-theory' programme, either of the compromises being anathematic to the very essence of GTR.

The quantized many-particle Hamiltonian representing a self-gravitating system has only recently been derived [4] as representing an exactly soluble sector of quantum gravity. Infact from the quantum gravity point of view after quantizing the GTR in the $c \to \infty$ (Newton-Cartan theory with spatial vanishing curvature), we have explicitly the above known Hamiltonian. We then special relativize the problem by using Mach's principle in case of the universe . Figure 1.1 represents the underling physical theories relating to full-blown quantum theory of gravity (denoted by FQG) which is still an elusive one. But ours is an effective theory of quantum gravity with vanishing spatial curvature ETQGK0 (denoted by a circle with squares inside) of this FQG with special-relativization done on the problem which is described by the quantum many-particle Hamiltonian obtained by quantizing the Newton-Cartan theory which we write as NQG (Fig. 1.1). Any general-relativistic exotic theory "theory of everything" (string theory, loop quantum gravity etc) would not be physically relevant if it does not reduce to Newton quantum gravity interacting with quantum fields in the Galilean-relativistic limit. Any valid theory of quantum gravity must reduce [4] to NQG (Fig. 1.1) in the "$c \to \infty$" limit, to GTR in the "$\hbar \to 0$" limit and

3

to QTF in the "$G \rightarrow 0$" limit. Since in our theory, we relativized the NQG, the elusive full-blown quantum theory of gravity must be consistent with the results obtained by our approach. The directions of research have been to go from GTR to FQG or from QTF to FQG. But the third possibility of going from NQG/QNG to FQG by special relativizing the Newton-Cartan quantum gravity (undoing the $c \rightarrow \infty$ limit) opens up an exciting yet unexplored direction in the research of quantum gravity.

We apply our theory for the study of quantum cosmology. Our expanding universe is made up of gravitationally interacting particles which are described by particles of luminous matter, dark matter and dark energy. Representing dark energy by a repulsive harmonic potential among the points in the flat 3-space, we derive a quantum mechanical relation connecting, temperature of the cosmic microwave background radiation, age, and cosmological constant of the universe. When the cosmological constant is zero, we get back Gamow's relation with a much better coefficient. Otherwise, our theory predicts a value of the cosmological constant 2.0×10^{-56} cm^{-2} when the present values of cosmic microwave background temperature of 2.728 K and age of the universe 14 billion years are taken as input. We study [3, 7, 8] the self-gravitating system such as the universe using a well-known many-particle Hamiltonian, which is known from the early days of quantum mechanics, from a condensed matter point of view by using a quantum mechanical variational approach. This can also be viewed as a novel way of looking at the self-gravitating systems and it not only reproduces the results known from Einstein's General Theory of Relativity but also goes beyond by predicting certain relations and specifically the value of the cosmological constant. We also study other self-gravitating systems like stars, neutron stars, white dwarfs etc using the Hamiltonian discussed above.

1.2 Goals of this book

In this book we establish an effective theory of quantum gravity and apply this theory to study self-gravitating systems such as stars, white dwarf, neutron star, black hole and also the universe through a quantum many-particle Hamiltonian. We use a variational approach to study these systems.

In this chapter 1, a general introduction to our effective theory of quantum gravity is already given. Also its relation to other theories of physics including the full blown quantum theory of gravity is depicted through a beautiful figure known as cube of theories.

In chapter 2, the generally-covariant but Galilean-relativistic theory of gravity with a possible non-zero cosmological constant, viewed as a parameterized gauge theory of a gravitational vector-potential minimally coupled to a complex Schrödinger-field (bosonic or fermionic), is successfully cast into a manifestly covariant Lagrangian form. This Newton-Cartan-Schrödinger system is non-perturbatively quantized. Consequently, the resulting theory of quantized Newton-Cartan-Schrödinger system constitutes a perfectly consistent Galilean-relativistic sector of the elusive full quantum theory of gravity coupled to relativistic matter. It turns out that the quantized N-C-S system is described by a Hamiltonian which is that of a self-gravitating system.

In chapter 3 we give an introduction to the relativization of quantized Newton-Cartan-Schrödinger system and its application to self-gravitating systems such as stars, black hole, and our expanding universe.

In chapter 4, we study the system of neutron stars and white dwarfs. We derive upper mass limit of a neutron star and the Chandrasekhar limit for the mass of a white dwarf in a novel way. We also obtain many other parameters of stars from our formalism.

In chapter 5, we derive the Schwarzschild radius of a black hole quantum mechanically and also find a quantum correction to this radius.

In chapter 6, we apply our formalism to the study of the universe and obtain upper limit on the mass of neutrino, number of nucleons in the universe, size and total mass of the universe.

In chapter 7, we study the expanding universe with the acceleration produced by the cosmological constant and derive a relation connecting time, temperature of the cosmic background radiation and the cosmological constant. We obtain a reasonably well value of the cosmological constant. Gamow's relation is revisited as a special case of the relation we have derived, when cosmological constant is zero.

In chapter 8, we give a summary and conclusion of our work. In appendix A, we derive different energy terms for a many-particle system, in appendix B we give simple description of the scale factor of the universe according to general theory of relativity . We give a short note on the principle of extremal action in appendix C and values of some important constants of nature in appendix D.

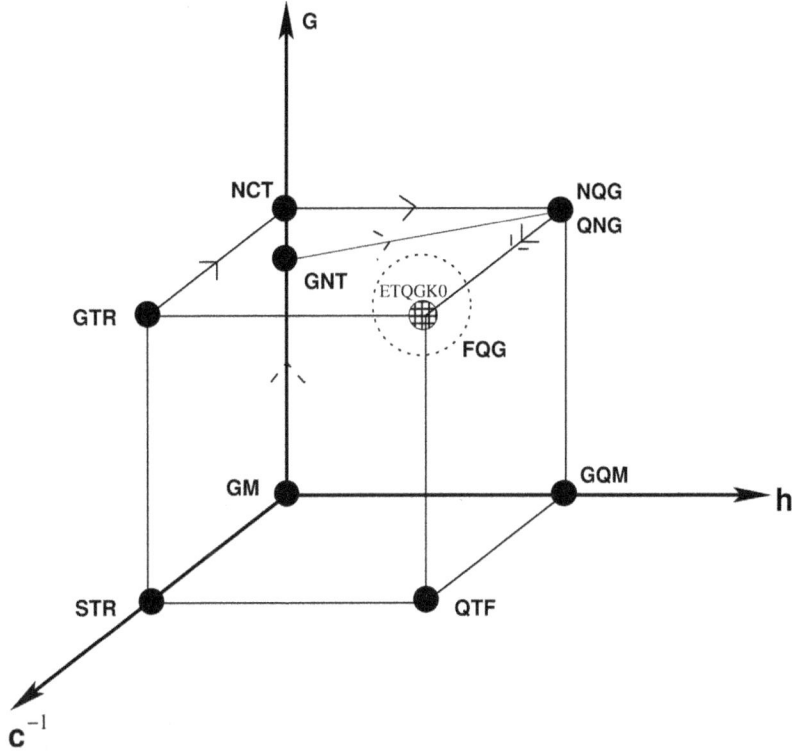

Figure 1.1: The great dimensional cube of physics indicating the fundamental roles played by G, h and c in various basic physical theories. The theories are GM=Galilean mechanics, STR=special theory of relativity, GTR=general theory of relativity, GQM=Galilean quantum mechanics, QTF=quantum theory of relativistic fields, NCT=Newtonian-Cartan theory, GNT=Galilean Newtonian theory, NQG=Newtonian quantum gravity, QNG=quantum Newtonian gravity. It turns out that when the NCT with (flat 3-space) is quantized (NQG), the resulting quantum many-particle Hamiltonian is same as one would write the quantum many-particle Hamiltonian for a system of self-gravitating particles that is what we denote as QNG. Here FQG=The elusive full-blown quantum gravity represented by the bigger circle with broken lines. The position denoted by a circle with squares inside denotes our effective theory of quantum gravity (ETQGK0) in the limiting case when special relativization is incorporated but the spatial curvature is zero, ETQGK0⊂ FQG. One can note two tracks to this limiting theory one shown by solid-arrow-heads starting from GTR and other by broken-arrow-heads from GM. The figure has been adopted and modified from Ref. [4].

Bibliography

[1] C. Rovelli, arXiv:gr-qc/0006061v3.

[2] C. Rovelli, *Quantum Gravity* (Cambridge University Press, Cambridge, England, 2004).

[3] S. Mishra, Int. J. Theor. Phys. **47**, 2655 (2008).

[4] J. Christian, Phys. Rev. D **56**, 4844 (1997).

[5] C. Kiefer, in *Beyond the Big Bang*, edited by R. Vaas (Springer, Berlin, 2008).

[6] J. Christian, in *Physics Meets Philosophy at the Planck scale*, edited by C. Callender and N. Huggett (Cambridge University Press, Cambridge, England, 2001).

[7] D. N. Tripathy and S. Mishra, Int. J. Mod. Phys. D **7**, 6, 917 (1998).

[8] D. N. Tripathy and S. Mishra, Int. J. Mod. Phys. D **7**, 3 , 431 (1998).

Part II

Quantizing the Hamiltonian :Exactly Soluble Sector of Quantum Gravity

Chapter 2

Quantized GTR when $c \to \infty$

Cartan's spacetime reformulation of the Newtonian theory of gravity is a generally-covariant Galilean-relativistic limit-form of Einstein's theory, degenerate 'metric' structure of spacetime remains fixed with two mutually orthogonal non-dynamical metrics, one spatial and the other temporal. The *spacetime* according to this theory is, nevertheless, *curved*, duly respecting the principle of equivalence, and the non-metric gravitational connection-field is *dynamical* in the sense that it is determined by matter distributions. Here, this generally-covariant but Galilean-relativistic theory of gravity with a possible non-zero cosmological constant, viewed as a parameterized gauge theory of a gravitational vector-potential minimally coupled to a complex Schrödinger-field (bosonic or fermionic), is successfully cast — for the first time — into a manifestly covariant Lagrangian form. Then, exploiting the fact that Newton-Cartan spacetime is intrinsically globally-hyperbolic with a fixed causal structure, the theory is recast both into a *constraint-free* Hamiltonian form in 3+1-dimensions and into a manifestly covariant reduced phase-space form with *non-degenerate* symplectic structure in 4-dimensions. Next, this Newton-Cartan-Schrödinger system is non-perturbatively quantized using the standard C*-algebraic technique combined with the geometric procedure of manifestly covariant phase-space quantization. The ensuing unitary quantum field theory of Newtonian gravity coupled to Galilean-relativistic matter is not only generally-covariant, but also *exactly soluble* and — thanks to the immutable causal structure of the Newton-Cartan spacetime — free of all conceptual and mathematical difficulties usually encountered in quantizing Einstein's theory of gravity. Consequently, the resulting theory of quantized Newton-Cartan-Schrödinger system constitutes a perfectly consistent Galilean-relativistic sector of the elusive full quantum theory of gravity coupled to relativistic matter, regardless of what ultimate form the latter theory eventually takes.

2.1 Introduction

The primary aim of this paper is to demonstrate that the principle of equivalence by itself is not responsible for the conceptual and mathematical difficulties encountered in constructing a viable quantum theory of gravity; rather, it is the conjunction of this principle with the conformal structure (i.e., the field of light-cones) of the general-relativistic spacetime which resists subjugation to the otherwise well-corroborated canonical rules of quantization. This elementary fact, of course, has been duly appreciated by the workers in the field at least implicitly since the earliest days of attempts to quantize Einstein's theory of gravity. Curiously enough, however, except for a partial illustration of this state of affairs by Kuchař [1], so far the problem of explicitly constructing a generally-covariant but Galilean-relativistic quantum theory of gravity — i.e., a generally-covariant quantum field theory of spacetime with degenerate structure of 'flattened' light-cones but unique *dynamical*[1] connection has been completely neglected [2].

(*1. The adjective 'dynamical' here and below simply refers to the mutability of spacetime structure dictated by evolving distributions of matter. It only refers to the fact that, even in a Galilean-relativistic theory, spacetime is not fixed a priori. In particular, unlike the case in general relativity, it does not refer to any*

11

transverse propagation degrees of freedom associated with the spacetime structure since nonrelativistic gravity does not possess such a freedom. See subsection 2.5.4, however, for a discussion on the longitudinal degrees of freedom of the gravitational field.)

Here we set out to construct such a 'nonrelativistic' theory and explicitly demonstrate that, unlike the case of relativistic quantum gravity, there are no insurmountable conceptual or mathematical difficulties in achieving this goal. In particular, we show that the Galilean-relativistic limit-form (see Figure 1) of the as-yet-untamed full quantum theory of gravity interacting with matter is *exactly soluble*, and that in this very classical ('c = ∞') domain 'the problem of time' [3] — the well-known central stumbling block encountered in quantizing Einstein's gravity — and related problems of causality, along with other conceptual and mathematical difficulties, are nonexistent.

The nonrelativistic limit-form of Einstein's theory of gravity is a spacetime reformulation of the Newtonian theory of gravity — the so-called Newton-Cartan theory. With the hindsight of Einstein's theory it is clear that gravitation should be treated as a consequence of the curving of spacetime rather than as a force-field even at the Newtonian level because the principle of equivalence is equally compatible with the Newtonian spacetime. A geometrical description of the Newtonian spacetime explicitly incorporating the principle of equivalence at the classical (non-quantal) level was given by Cartan [4] and Friedrichs [5] soon after the completion of Einstein's theory, and later further developed by many authors [6–10]. The outcome of such a reformulation of the Newtonian theory is a theory whose qualitative features lie in-between those of special relativity with its completely fixed spacetime background and general relativity with no background structure whatsoever. Unlike the latter two well-known theories, Newton-Cartan theory has two fixed and degenerate metrics — a temporal metric and a spatial metric, vaguely resembling the fixed Minkowski metric as far as their non-dynamical character is concerned — and a non-metric but *dynamical*[1] connection-field mimicking Einstein's metric connection-field to some extent. Thus, unlike the static and flat Galilean spacetime, and analogous to the mutable general-relativistic spacetime, the generally-covariant[2] Newton-Cartan spacetime is dynamical, curved by the Newtonian gravity, and requires no *a priori* assumption of a global inertial frame.

(2. The philosophical dispute over the meaning of the principle of general covariance begun almost immediately after the completion of Einstein's theory of gravity [14] and persists today [15]. Einstein, for instance, read through Stachel's eyeglasses [16], would not view Newton-Cartan theory as a genuinely generally-covariant theory because, unlike general relativity, it does not avert the 'hole argument' [17]. As far as this paper is concerned, however, we can afford to refrain from the controversy and join the bandwagon calling Newton-Cartan spacetime generally-covariant simply because it respects the principle of equivalence and precludes existence of global inertial frames of reference. Moreover, at least in the framework followed in this paper, the non-metric Newton-Cartan theory is as diffeomorphism-invariant as Einstein's metric theory of gravity (cf. Eq. (2.50). (See also note 5.))

Consequently, the transition from Galilean spacetime to this general-*non*relativistic Newton-Cartan spacetime drastically changes the *qualitative* features of the Galilean-relativistic physics by elevating the status of the affine connection from that of an absolute element — given once and for all — to a dynamical quantity determined by the distribution of matter. Furthermore, it is this Newton-Cartan theory of gravity with its mutable spacetime which is in general the true Galilean-relativistic limit-form of Einstein's general theory of relativity [10–13], and not the Newtonian theory on the immutable background of flat

Galilean spacetime (cf. subsection 2.2.3 below). In summary, Cartan's geometric reformulation of Newton's theory of gravity makes it a *local* field theory analogous to general relativity, and, as a result, the instantaneous gravitational interactions between gravitating bodies can now be understood as propagating continuously through the curvature of the region of spacetime among them.

To understand these stipulations in detail, in section 2.2 we accumulate various scattered results to provide a coherent — but by no means exhaustive — review of the classical (non-quantal) Newton-Cartan theory of gravity. Then, in section 2.3, we review the previous work on one-particle Schrödinger quantum mechanics on the curved Newton-Cartan spacetime initiated by Kuchař [1], and extend it to a Galilean-relativistic quantum field theory on such a spacetime. Next, in section 2.4, we derive, *for the first time*, the complete classical Newton-Cartan-Schrödinger theory (i.e., the theory of classical Newton-Cartan field interacting with a Galilean-relativistic matter) from extremizations of a single diffeomorphism-invariant action functional. This functional — defined on an arbitrary measurable region of spacetime — is carefully selected to allow recasting of the theory into a constraint-free Hamiltonian form in 3+1 dimensions, as well as into a manifestly covariant reduced phase-space form in 4-dimensions (where the phase-space is viewed as the space of solutions of the equations of motion modulo gauge-transformations). Thus obtained covariant and constraint-free phase-space then paves the way in section 2.5 for a straightforward quantization of the Newton-Cartan-Schrödinger system using the standard C*-algebraic techniques and the associated representation theory. The resulting local quantum field theory describes an *exactly* soluble interacting matter-gravity system.

In addition to the primary aim discussed in the beginning of this Introduction, there are various other motivations for the present exercise which we enumerate here (counting the primary aim as (1)):

14

(2) As we shall see in the next section, one of the set of gravitational field equations of Newton-Cartan theory closely resembles Einstein's field equation $G_{\mu\nu} = 8\pi T_{\mu\nu}$ relating geometry to matter. In particular, it dictates that the connection-field of Newton-Cartan gravity must be determined by the distribution of matter. Therefore, consistency requires that this connection-field must be quantized along with matter since it participates in the dynamical unfolding of the combined system.

(3) It is well-known that Newtonian models play quite a significant role in cosmology. In particular, they are useful in studying structure formation in the early universe [18] and provide useful insights for the relativistic case. Recently, the need to generalize Newtonian cosmology to Newton-Cartan cosmology has been recognized, and a considerable progress has been made in this direction [19–21]. In this context, then, the relevance of an exactly soluble quantum theory of Newton-Cartan gravity interacting with matter cannot be overestimated.

(4) It is fair to say that we know very little about the quantum gravity proper (i.e., FQG in the Figure 1). Therefore, insights coming form any quarters which enable us to better understand the difficulties of constructing the final theory should be welcomed. It is with this attitude that the recent exercises on exactly soluble gravitating systems in reduced spacetime dimensions [22] and/or reduced symmetries [23] have been carried out, and used as probes to investigate the conceptual problems of the full quantum gravity. Here we do not reduce either symmetries or the spacetime dimensions, but instead provide an exactly soluble system in the full 4-dimensional setting — albeit only in the Galilean-relativistic limit of the full theory.

(5) In Ref. [24] we have argued, on both group theoretical and physical grounds, that, since Newton-Cartan symmetries — duly respecting the principle of equivalence — and not Galilean symmetries are the true spacetime symmetries

of the nonrelativistic quantum domain, any discussion on the conceptual issues like the 'measurement problem' in this domain must be carried out within the Newton-Cartan framework. In fact, it was argued, Penrose-type speculations of gravitationally induced state-reduction [25] might greatly benefit from analyzing the relevant physical systems within this framework. It is gratifying to note that this suggestion has already attracted at least Penrose's attention [25, 26]. However, the complete framework for such deep conceptual issues must be the fully quantized Newton-Cartan gravity interacting with matter — and hence the present work.

(6) The theory constructed in this paper provides a selection criterion for any exotic, top-down approach (e.g., the superstring approach) to the final 'theory of everything.' Clearly, any general-relativistic exotic theory would lose its physical relevance if it does not reduce to the Newtonian quantum gravity interacting with Schrödinger-fields in the Galilean-relativistic limit. Therefore, any future theory must reduce to NQG of Figure 1 in the '$c \to \infty$' limit, as much as it should reduce to GTR in the '$\hbar \to 0$' limit and QTF in the '$G \to 0$' limit.

(7) Finally, the existence of NQG opens up a completely novel direction of research in the full quantum gravity. In majority of orthodox approaches to quantum gravity the direction of research has been to go from GTR to FQG (cf. Figure 1) — i.e., the program has been to quantize the general theory of relativity. Somewhat less popular and less explored program is to go from QTF to FQG — i.e., to general-relativize the quantum theory of fields [27]. The existence of NQG opens up a third possibility, that of starting from NQG and arriving at FQG by undoing the '$c \to \infty$' limit — i.e., by special-relativizing the Newton-Cartan quantum gravity.

2.2 Spacetime approach to Newtonian gravity

In this section we review the covariant, spacetime reformulation of the Newtonian theory of gravity, and, thereby, set the notations and conventions to be used in the following sections. Most of the ideas presented in this and the next section are not new, but it is the manner in which they are organized here that makes them conducive to the fruitful results of the later sections.

2.2.1 General Galilean spacetime

We begin with the familiar spacetime structure $(\mathcal{M}; t_\alpha, h^{\alpha\beta}, \nabla_\alpha)$ presupposed by the usual Galilean-relativistic dynamics delineated in Penrose's abstract index notation [28] using the greek alphabet [4, 9, 13]. *Spacetime* — the arena in which physical events and processes take place — is represented by a real, contractible, and differentiable Hausdorff 4-manifold $\mathcal{M}$ without boundaries. Unlike the case in general-relativistic spacetimes, here spatial and temporal measures on $\mathcal{M}$ are not taken to be soldered into a single semi-Riemannian metric, but appear as two distinct geometric entities. A smooth, never vanishing covariant vector field t_α on $\mathcal{M}$ is defined which induces a degenerate *temporal metric* $t_{\mu\nu} = t_{(\mu\nu)} := t_\mu t_\nu$ of signature $(+\,0\,0\,0)$ specifying durations of processes occurring between events, and hence induces a degenerate 'cone structure' on the tangent space $T_x\mathcal{M}$ at each point $x \in \mathcal{M}$ (here the parentheses indicate symmetrization with respect to the enclosed indices). A vector ξ^μ at a point on $\mathcal{M}$ is said to be *time-like* if $\sqrt{t_{\mu\nu}\xi^\mu\xi^\nu} > 0$, *spacelike* if $\sqrt{t_{\mu\nu}\xi^\mu\xi^\nu} = 0$, and *future-directed* if $t_\mu\xi^\mu > 0$. Between spacelike vectors ξ^α on $\mathcal{M}$ with vanishing 'temporal lengths' — i.e., $\sqrt{t_{\mu\nu}\xi^\mu\xi^\nu} = 0$ — there is defined an inner product with the help of a smooth, symmetric, contravariant vector field $h^{\alpha\beta} = h^{(\alpha\beta)}$ on $\mathcal{M}$ which serves as a degenerate *spatial metric* of signature $(0 + + +)$, and indirectly assigns lengths to these vectors: $|\xi| := \sqrt{h^{\mu\nu}\lambda_\mu\lambda_\nu}$, where $h^{\mu\nu}\lambda_\nu = \xi^\mu$ with an arbitrary choice of λ_μ. As we have done here, the metric tensor $h^{\mu\nu}$ can be used to raise indices; however, since it is not invertible, it cannot be used to lower indices. Thus, the distinction between covectors and contravectors has much greater significance in this general Galilean-relativistic spacetime than in the semi-Riemannian general-relativistic spacetime. The *affine structure* of the spacetime $\mathcal{M}$ is represented by a smooth derivative operator ∇_α introducing a (not necessarily 'flat') torsion-free linear connection Γ on $\mathcal{M}$. The two tensor fields — $\tau := t_\alpha dx^\alpha$ measuring the proper time of world lines and $h := h^{\alpha\beta}\partial_\alpha \otimes \partial_\beta$ inducing a 3-metric on the null space of τ — are mutually orthogonal,

$$\tau \lrcorner h \,=\, 0 \,=\, h^{\alpha\beta}t_\beta \tag{2.1}$$

(the kernel of h generating the span of the 1-form τ), and taken to be compatible with the derivative operator:

$$\nabla_\alpha h^{\beta\gamma} = 0 = \nabla_\alpha t_\beta \,. \tag{2.2}$$

A Galilean spacetime is orientable as there exists a 4-volume element for the structure $(\mathcal{M};\, t_\alpha,\, h^{\alpha\beta},\, \nabla_\alpha)$. Given the structure $(\mathcal{M};\, t_\alpha,\, h^{\alpha\beta},\, \nabla_\alpha)$ satisfying the conditions 2.2, a continuous, nowhere vanishing spacetime measure form on $\mathcal{M}$ with tensor components $\varepsilon_{\alpha\beta\gamma\delta} = \varepsilon_{[\alpha\beta\gamma\delta]}$ can be derived [29] such that $\nabla_\mu \varepsilon_{\alpha\beta\gamma\delta} = 0$ (where the square brackets indicate anti-symmetrization with respect to the bracketed indices). If one defines an oriented *Galilean frame* at a point $x \in \mathcal{M}$ as a basis $\{e_i\}$ of the tangent space $T_x\mathcal{M}$ such that $e_i^\alpha t_\alpha = \delta_i^0$ and $h^{\mu\nu}\theta_\mu^i\theta_\nu^j = \delta_a^i \delta^{ab}\delta_b^j$ (where $\alpha, i, j = 0,...,3$ and $a, b = 1, 2, 3$), with $\{\theta^i\}$ being the dual basis of $\{e_i\}$, then for any such Galilean frame $\{e_i\}$ the canonical 4-volume element can be defined by [11]

$$\wp\, d^4x \,:=\, \frac{1}{4!}\, \varepsilon_{\alpha\beta\gamma\delta}\, dx^\alpha \wedge dx^\beta \wedge dx^\gamma \wedge dx^\delta \,:=\, \theta^0 \wedge \theta^1 \wedge \theta^2 \wedge \theta^3 \,. \tag{2.3}$$

The compatibility of the temporal metric $\nabla_\alpha t_\beta = 0$, giving the condition $\nabla_{[\gamma} t_{\delta]} = 0$, at least locally allows the relation $t_\beta = \nabla_\beta t$ for some time function t. Since $\mathcal{M}$ is contractible by definition, the Poincaré lemma allows one to define the absolute time function also globally by a map $t : \mathcal{M} \to \mathbb{R}$, foliating the spacetime *uniquely* into one-parameter family of (not necessarily flat) spacelike hypersurfaces[3] of simultaneity. One can use this time-function t as an affine parameter for arbitrary timelike curves representing the worldlines of test particles.

(3. As is well-known, hypersurfaces are best represented by embeddings. A one-parameter family of embeddings in the present context is a map $^{(t)}e : \Sigma \to \mathcal{M}$ which takes a point x^a from the spatial submanifold Σ of $\mathcal{M}$ to a point $x^\alpha(x^a, t)$ in the spacetime, where $t \in \mathbb{R}$ labels the leaves of this foliation. A hypersurface, then, is an equivalence class of such embeddings modulo diffeomorphisms of the submanifold Σ (cf. subsection 2.4.2).)

Since the compatibility condition $\nabla_\mu t_\nu = 0$ leads to the relation $t_\gamma \Gamma^\gamma_{[\mu \nu]} = \partial_{[\mu} t_{\nu]}$, it is clear that a torsion-free connection,

$$\Gamma^\gamma_{[\mu \nu]} = 0 , \tag{2.4}$$

is admissible only if

$$\partial_{[\mu} t_{\nu]} = 0 ; \quad \text{i.e., } d\tau = 0 . \tag{2.5}$$

In fact, it can be shown [9] that the closed-ness of τ is both necessary and sufficient condition for the existence of a torsion-free connection as a part of the Galilean structure. Moreover, without the condition that the 1-form τ is closed, the structure $(\mathcal{M}; \tau, h)$ is *not integrable*. A linear symmetric Galilean connection satisfying the compatibility conditions (2.2) exists if and only if the structure $(\mathcal{M}; \tau, h)$ is integrable. However, given integrability, such a symmetric connection is unique *only* up to an arbitrary 2-form $F = \frac{1}{2} F_{\mu\nu} \, dx^\mu \wedge dx^\nu$, and can be decomposed as [30]:

$$\Gamma_{\mu\ \nu}^{\ \gamma} \ = \ \Gamma_{(\mu\ \nu)}^{\ \ \gamma} \ = \ \overset{u}{\Gamma}_{\mu\ \nu}^{\ \gamma} + t_{(\mu} F_{\nu)\sigma} h^{\sigma\gamma} , \qquad (2.6)$$

where [32]

$$\overset{u}{\Gamma}_{\mu\ \nu}^{\ \gamma} := \ h^{\gamma\sigma} \{ \partial_{(\mu} \overset{u}{h}_{\nu)\sigma} - \tfrac{1}{2}\partial_\sigma \overset{u}{h}_{\mu\nu} \} + u^\gamma \partial_{(\mu} t_{\nu)} \qquad (2.7)$$

with $u = u^\alpha \partial_\alpha$ representing an arbitrary unit timelike vector-field,

$$t_\alpha u^\alpha \ = \ 1 \ = \ \tau \lrcorner\, u , \qquad (2.8)$$

interpreted as the four-velocity of an adscititious 'æther-frame' and $\overset{u}{h}_{\mu\nu} = \overset{u}{h}_{(\mu\nu)}$ representing the associated spatial projection field *relative to* u^α defined by

$$\overset{u}{h}_{\mu\rho} u^\rho := 0 \quad \text{and} \quad \overset{u}{h}_{\mu\rho} h^{\rho\nu} := \delta_\mu^{\ \nu} - t_\mu u^\nu =: \overset{u}{\delta}_\mu^{\ \nu} . \qquad (2.9)$$

Throughout the paper, a letter (e.g., u) on the top of a quantity indicates gauge-dependence of the quantity (e.g., dependence of the quantity $\overset{u}{h}_{\mu\nu}$ on the 'gauge' u). This relative projection field $\overset{u}{h}_{\mu\nu}$ introducing a fiducial 'absolute rest' [13] may be used to lower indices, but, of course, only relative to u; for, under an æther-frame transformation of the form

$$u^\alpha \ \mapsto \ \tilde{u}^\alpha \ := \ u^\alpha \ + \ h^{\alpha\sigma} \mathrm{w}_\sigma \qquad (2.10)$$

for a boost covector w_σ , the relative projection field undergoes a nontrivial boost transformation:

$$\overset{u}{h}_{\mu\nu} \ \mapsto \ \overset{\tilde{u}}{h}_{\mu\nu} \ := \ \overset{u}{h}_{\mu\nu} \ - \ 2\, t_{(\mu} \overset{u}{h}_{\nu)\alpha} \, h^{\alpha\sigma} \mathrm{w}_\sigma \ + \ t_{\mu\nu} \, h^{\alpha\sigma} \mathrm{w}_\alpha \mathrm{w}_\sigma . \qquad (2.11)$$

In other words,

$$\frac{\partial}{\partial u^\sigma} \overset{u}{h}_{\mu\nu} \ = \ - \, 2\, t_{(\mu} \overset{u}{h}_{\nu)\alpha} , \qquad (2.12)$$

which follows from the definition (2.9).

20

The special connection $\overset{u}{\Gamma}$ is unique and symmetric, and such that the Galilean observer u^α associated with it is geodetic, $u^\sigma \overset{u}{\nabla}_\sigma u^\alpha = 0$, and curl-free, $h^{\sigma[\mu} \overset{u}{\nabla}_\sigma u^{\nu]} = 0$. Using these two properties of u^α, and the above definition of its relative projection field $\overset{u}{h}_{\mu\nu}$, one can express the 2-form F in terms of the covariant derivative of this observer u^α [30]:

$$F_{\mu\nu} = -2 \, \overset{u}{h}_{\sigma[\mu} \nabla_{\nu]} u^\sigma . \tag{2.13}$$

For future purposes, we also observe that the spatial projection field $\overset{u}{h}_{\mu\nu}$ is *not* compatible with the covariant derivative operator ∇_σ in general, but, instead, its covariant derivative depends on the covariant rate of change of the observer-field u^α [30]:

$$\nabla_\sigma \overset{u}{h}_{\mu\nu} = -2 \{ \nabla_\sigma u^\alpha \} \overset{u}{h}_{\alpha(\mu} t_{\nu)} , \tag{2.14}$$

where the covariant derivative of u^α may be decomposed as [31]

$$\nabla_\mu u^\nu = \overset{u}{h}_{\mu\alpha} h^{\sigma(\alpha} \nabla_\sigma u^{\nu)} + \overset{u}{h}_{\mu\alpha} h^{\sigma[\alpha} \nabla_\sigma u^{\nu]} + t_\mu u^\sigma \nabla_\sigma u^\nu . \tag{2.15}$$

The first term of this decomposition is a measure of the lack of 'rigidity' of the field u^α, and, hence, in the case of the rigid Galilean frame it vanishes identically: $h^{\sigma(\mu} \nabla_\sigma u^{\nu)} \equiv 0$. Further, since the Galilean observer u^α is curl-free and geodetic with respect to the associated connection $\overset{u}{\Gamma}$, the remaining two terms of the Eq. (2.15) also vanish in this special case giving the uniformity property

$$\overset{u}{\nabla}_\mu u^\nu = 0 ; \tag{2.16}$$

i.e., u^ν is covariantly constant with respect to the special connection $\overset{u}{\Gamma}$. As a result of this property, Eq.(2.14) yields the compatibility relation for the relative spatial projection field $\overset{u}{h}_{\mu\nu}$,

$$\overset{u}{\nabla}_\sigma \overset{u}{h}_{\mu\nu} = 0 , \tag{2.17}$$

in this special case of unique derivative operator $\overset{u}{\nabla}_\sigma$ associated with the æther-frame u^α.

The curvature tensor corresponding to the symmetric Galilean connection (2.6) —defined by $R^\sigma{}_{\beta\gamma\delta} f_\sigma = 2\nabla_{[\gamma}\nabla_{\delta]} f_\beta$, or, by $R^\alpha{}_{\sigma\gamma\delta} g^\sigma = -2\nabla_{[\gamma}\nabla_{\delta]} g^\alpha$, for arbitrary vectors f_α and g^α — observes the symmetries [9, 30]

$$h^{\sigma(\beta} R^{\alpha)}{}_{\sigma\gamma\delta} = 0 \quad \text{and} \quad t_\sigma R^\sigma{}_{\beta\gamma\delta} = 0 \tag{2.18}$$

by virtue of the metric compatibility conditions, in addition to the usual constraints

$$R^\alpha{}_{\beta(\gamma\delta)} = 0, \quad R^\alpha{}_{[\beta\gamma\delta]} = 0, \quad \text{and} \quad R^\alpha{}_{\beta[\gamma\delta;\lambda]} = 0 \tag{2.19}$$

satisfied by any curvature tensor, where $;\lambda$ denotes the covariant derivative ∇_λ. It can be easily verified that the contracted Bianchi identities of the form

$$\nabla_\nu(R^{\mu\nu} - \tfrac{1}{2} R h^{\mu\nu}) = 0 \tag{2.20}$$

hold in the general Galilean spacetime as a consequence of the last of these constraints, where $R_{(\mu\nu)} = R_{\mu\nu} := R^\sigma{}_{\mu\nu\sigma}$ and $R := h^{\mu\nu} R_{\mu\nu}$ are the corresponding Ricci tensor and Ricci scalar, respectively.

Finally, we close this subsection by reemphasizing that, although an integrable Galilean structure $(\mathcal{M}; t_\alpha, h^{\alpha\beta}, \nabla_\alpha)$ is completely specified by the four conditions

$$h^{\alpha\beta} t_\beta = 0, \quad \nabla_\alpha h^{\beta\gamma} = 0, \quad \nabla_\alpha t_\beta = 0, \quad \text{and} \quad \partial_{[\alpha} t_{\beta]} = 0, \tag{2.21}$$

the Galilean connection (2.6) remains under-determined, in general, by an arbitrary 2-form.

2.2.2 Specializing to the Newton-Cartan spacetime

Now, Cartan's spacetime reformulation of the Newtonian theory of gravity can be motivated in exact analogy with Einstein's theory of gravity. The analogy works because the universal equality of the inertial and the passive gravitational masses is independent of the relativization of time, and hence is equally valid at the Galilean-relativistic level. As a result, it is possible to parallel Einstein's theory and reconstrue the trajectories of (only) gravitationally affected particles as geodesics of a unique, 'non-flat' connection Γ satisfying

$$a^i := \frac{d^2 x^i}{dt^2} + \Gamma_{j\,k}^{\ i} \frac{dx^j}{dt} \frac{dx^k}{dt} = 0 \qquad (2.22)$$

in a coordinate basis, or, equivalently,

$$a^\alpha := v^\sigma \nabla_\sigma v^\alpha = 0 \qquad (2.23)$$

in general, such that

$$\Gamma_{\nu\ \lambda}^{\ \mu} \equiv \overset{v}{\Gamma}_{\nu\ \lambda}^{\ \mu} + \overset{v}{G}_{\nu\ \lambda}^{\ \mu} := \overset{v}{\Gamma}_{\nu\ \lambda}^{\ \mu} + h^{\mu\alpha} \overset{v}{\nabla}_\alpha \overset{v}{\Phi}\, t_{\nu\lambda} \qquad (2.24)$$

with $\overset{v}{\Phi}$ representing the Newtonian gravitational potential relative to the freely falling observer field v, $\overset{v}{\Gamma}_{\nu\ \lambda}^{\ \mu}$ representing the coefficients of the corresponding 'flat' connection (i.e., one whose coefficients can be made to vanish in a suitably chosen linear coordinate system), and

$$\overset{v}{G}_{\nu\ \lambda}^{\ \mu} := h^{\mu\alpha} \overset{v}{\nabla}_\alpha \overset{v}{\Phi}\, t_{\nu\lambda} \qquad (2.25)$$

representing the traceless gravitational field tensor associated with the Newtonian potential [29]. The conceptual superiority of this geometrization of Newtonian gravity is reflected in the trading of the two 'gauge-dependent' quantities $\overset{v}{\Gamma}$ and $\overset{v}{G}$ in favor of their gauge-independent sum Γ. Physically, it is the 'curved' connection Γ rather than any 'flat' connection $\overset{v}{\Gamma}$ that can be determined by local experiments. The potential $\overset{v}{\Phi}$ and the 'flat' connection $\overset{v}{\Gamma}$ do not have an independent existence; they exist only relative to an arbitrary choice of inertial frame. Given the 'curved' connection Γ, its associated curvature tensor works out to be

$$R^{\alpha}{}_{\beta\gamma\delta} \;=\; 2\,t_{\beta}\,h^{\alpha\lambda}\,\overset{v}{\Phi}_{;\lambda;[\gamma}\,t_{\delta]}\;,\qquad\qquad(2.26)$$

where $;\alpha$ represent covariant derivatives $\overset{v}{\nabla}_{\alpha}$ corresponding to the 'flat' connection $\overset{v}{\Gamma}$. Although written in terms of gauge-dependent quantities, this expression for the curvature tensor is, of course, gauge-invariant.

The non-flat connection Γ in equation (2.22) is still compatible with the temporal and spatial metrics τ and h: $\nabla_{\alpha}t_{\beta} = \nabla_{\alpha}h^{\beta\gamma} = 0$. This non-uniqueness of the compatible connections is due to the degenerate nature of the metrics τ and h — i.e., due to the non-semi-Riemannian nature of the Galilean structure $(\mathcal{M};\,\tau,\,h,\,\nabla)$. Unlike in the semi-Riemannian spacetime structures of the special and general theories of relativity, here the covariant derivative operator ∇_{α} is not fully determined; as noted in the previous subsection, the connection (2.6) is determined by the compatibility conditions (2.2) *only* up to an arbitrary 2-form. Therefore, in addition to the structure $(\mathcal{M};\,\tau,\,h\,)$, a connection must be specified to construct a completely geometrized, spacetime formulation of the Galilean-relativistic physics. In the case of Newtonian theory of gravity, geometrization can be easily achieved by taking those compatible connections Γ for the structure $(\mathcal{M};\,\tau,\,h,\,\nabla)$ whose curvature tensor, in addition to the properties (2.18) and (2.19), satisfies [9]

$$R^{\alpha}{}_{\beta}{}^{\gamma}{}_{\delta} = R^{\gamma}{}_{\delta}{}^{\alpha}{}_{\beta}\;,\qquad\qquad(2.27)$$

where $R^{\alpha}{}_{\beta}{}^{\gamma}{}_{\delta} \equiv h^{\gamma\sigma}R^{\alpha}{}_{\beta\sigma\delta}$. This extra condition roughly expresses the curl-freeness of the Newtonian gravitational field, and, together with the constraints 2.21, *uniquely* specifies the 'curved' Newton-Cartan connection; it is clearly satisfied by any Galilean connection that is obtained as a limit of the torsion-free Lorentzian connection. Equivalently, one can specify the Newton-Cartan connection by requiring that the 2-form F in equation (2.6) be closed [9, 30]: $dF = 0$. Then, in the light of the identity $d^2 = 0$, the 2-form F may be expressed in terms of an arbitrary 1-form A as

$$F_{\mu\nu} = 2\nabla_{[\mu}A_{\nu]} = 2\partial_{[\mu}A_{\nu]}. \tag{2.28}$$

Clearly, this expressibility of the 2-form F as an exterior derivative of an arbitrary 1-form is at least a *sufficient* condition for the connection (2.6) to be Newton-Cartan. Since $\mathcal{M}$ is contractible by definition, however, the Poincaré lemma implies that (2.28) is also a *necessary* condition for the specification of the Newton-Cartan connection. Comparison of equations (2.28) and (2.13) shows that, at least locally, the intrinsic condition (2.27) can be equivalently expressed in terms of the fields u^ν and A_μ as [34]

$$\nabla_{[\mu}A_{\nu]} + \overset{u}{h}_{\sigma[\mu}\nabla_{\nu]}u^\sigma = 0. \tag{2.29}$$

It is this convenient local form of the condition which will be useful in latter sections.

Physically, the covariant vector A_μ may be interpreted as the 'vector-potential' representing the combination of gravitational and inertial effects with respect to the Galilean observer $u = u^\alpha\partial_\alpha$. Given the 'æther-frame' based observer u^α, an *arbitrary* observer may be characterized by a normalized four-velocity vector v^α, $v^\alpha t_\alpha = 1$, such that the difference vector $h^{\alpha\sigma}A_\sigma := u^\alpha - v^\alpha$ describes how the motion of this arbitrary observer deviates from that of the 'æther-frame'. In terms of u and A any Newton-Cartan connection can be affinely decomposed as

$$\Gamma^{\gamma}_{\alpha\,\beta} = \overset{u}{\Gamma}{}^{\gamma}_{\alpha\,\beta} + \overset{A}{\Gamma}{}^{\gamma}_{\alpha\,\beta}, \tag{2.30}$$

where

$$\overset{A}{\Gamma}{}^{\gamma}_{\alpha\,\beta} := t_{(\alpha}[\partial_{\beta)}A_\sigma - \partial_\sigma A_{\beta)}]h^{\sigma\gamma}. \tag{2.31}$$

By comparing this decomposition in terms of the 'gauge' (u, A) with that in terms of the generic 'gauge' $(v, \overset{v}{\Phi})$ (cf. equation (2.24)) we find the relation

$$
\begin{aligned}
v^\alpha &= u^\alpha - h^{\alpha\sigma} A_\sigma \\
\overset{v}{\Phi} &= \frac{1}{2} h^{\mu\nu} A_\mu A_\nu - A_\sigma u^\sigma
\end{aligned}
\tag{2.32}
$$

between the two gauges [33]. With $A^2/2$ being the 'Coriolis' rotational potential [1], we recognize that the observer v is in 4-rotation with respect to the æther u. Conversely, the vector-potential A may be expressed in terms of the observer field v and its relative Newtonian gravitational potential $\overset{v}{\Phi}$ as [33]

$$
A_\mu = -\overset{u}{h}_{\mu\alpha} v^\alpha + [\overset{v}{\Phi} - \frac{1}{2} \overset{u}{h}_{\nu\sigma} v^\nu v^\sigma] t_\mu .
\tag{2.33}
$$

We noted in the previous subsection that given a Galilean structure $(\mathcal{M}, h, \tau)$ and an observer field u, there exists an associated unique torsionless connection $\overset{u}{\Gamma}$ with respect to which the observer is geodetic and curl-free. It turns out that in the case of Newton-Cartan connection defined by (2.27), the converse is also true [8]: for every Newton-Cartan connection Γ, at least locally there exists a unit, timelike, nonrotating, and freely-falling observer u such that $\Gamma = \overset{u}{\Gamma}$. For such an observer, the uniformity property (2.16) immediately leads to

$$
u^\sigma R^\alpha{}_{\sigma\gamma\delta} = -2\nabla_{[\gamma}\nabla_{\delta]} u^\alpha = 0
\tag{2.34}
$$

implying that its parallel-transport around a small closed curve is path-independent. But, of course, for any other timelike observer $\tilde{u}^\sigma = u^\sigma + h^{\sigma\mu} w_\mu$, with w being an arbitrary 1-form, we would instead have

$$
\tilde{u}^\sigma R^\alpha{}_{\sigma\gamma\delta} = h^{\sigma\mu} w_\mu R^\alpha{}_{\sigma\gamma\delta} \equiv w_\mu R^{\alpha\mu}{}_{\gamma\delta} \neq 0,
\tag{2.35}
$$

in general, unless a further restriction, $h^{\mu\sigma} R^\alpha{}_{\sigma\gamma\delta} = 0$, is imposed on the curvature tensor. That is, parallel-transport of an arbitrary observer around a small closed curve is path-dependent in general unless $R^{\alpha\mu}{}_{\gamma\delta} = 0$ everywhere. This innocuous looking extra condition turns out to be conceptually quite significant as far as a clearer understanding of the limit-relation between Einstein's and Newton's theories of gravity is concerned. We shall devote the entire next subsection to elaborate on its significance.

As in the general theory of relativity, spacetime becomes *dynamical*[1] under this geometrical reformulation of Newton's theory: the affine structure of the spacetime — i.e., the connection Γ — crucially depends on the distribution of matter ρ, and participates in the unfolding of physics rather than being a passive backdrop for the unfolding. This mutability of the Newton-Cartan spacetime is captured in a generalized Newton-Poisson equation which dynamically correlates the curvature of spacetime with the presence of matter:

$$R_{\alpha\beta} = 4\pi G \, \rho \, t_{\alpha\beta} =: 4\pi G \, P_{\alpha\beta} \,, \tag{2.36}$$

where $R_{(\alpha\beta)} = R_{\alpha\beta} := R^{\gamma}{}_{\alpha\beta\gamma}$ is the Ricci tensor corresponding to the connection Γ, G is the Newtonian gravitational constant, and $P_{\mu\nu} = \rho \, t_{\mu\nu}$ are the tensor components of the momentum tensor defined on $\mathcal{M}$ (seen to be as such by writing $P^{\alpha\sigma} := \rho \, u^{\alpha} u^{\sigma}$ and then lowering the indices by $t_{\mu\nu}$: $P_{\mu\nu} = t_{\mu} t_{\alpha} t_{\nu} t_{\sigma} P^{\alpha\sigma}$). The constraint (2.36) on the Ricci tensor is a Galilean-relativistic limit form of Einstein's equation, and, unlike the possible foliation of the general Galilean spacetime in terms of arbitrary (i.e., possibly curved) spacelike hypersurfaces discussed in the previous subsection, its presence in this specialized Newton-Cartan spacetime necessitates that the hypersurfaces of simultaneity remain copies of ordinary Euclidean three-spaces: $h^{\mu\alpha} h^{\nu\beta} R_{\alpha\beta} = 0$ because $h^{\mu\nu} t_{\nu} = 0$ [13]. More generally $R^{\mu\nu}$ may be defined as

$$R^{\mu\nu} t_{\mu\alpha} t_{\nu\beta} := R_{\alpha\beta} \,, \tag{2.37}$$

subject to the consistency condition $\overset{u}{h}_{\mu\nu}R^{\mu\nu} := R = h^{\mu\nu}R_{\mu\nu}$.

In contrast to the general theory of relativity, here the mass density ρ is the only source of the gravitational field. However, without loss of consistency, *spacelike* tensor fields, say s^α and $S^{\alpha\sigma} = S^{(\alpha\sigma)}$, may be added to the matter side of the above field equation as long as the resulting mass-momentum-stress tensor (or matter tensor, for short)

$$M^{(\alpha\sigma)} = M^{\alpha\sigma} := P^{\alpha\sigma} - 2\,u^{(\alpha}s^{\sigma)} - S^{\alpha\sigma} \tag{2.38}$$

describing continuous matter distributions satisfies the conservation condition

$$\nabla_\alpha M^{\alpha\sigma} = 0. \tag{2.39}$$

As a matter of fact, expression (2.38) is the only consistent possible nonrelativistic limit-form of the relativistic stress-energy tensor [11]. Note that, unlike the case in general relativity, here this conservation law *must be postulated independently* to allow the derivation of Newtonian equations of motion (2.22) in a manner analogous to the derivation of Einsteinian equations of motion from the general relativistic conservation law. In other words, at least for now, the matter conservation condition (2.39) must be viewed as a separate field equation of the theory. As we shall see, however, it can be *derived* using only the principle of general covariance in a variational approach discussed in the subsection 2.2.5 below.

As in Einstein's theory, the field equation (2.36) is generalizable by an additional cosmological term:

$$R_{\mu\nu} + \Lambda\,t_{\mu\nu} = 4\pi G\,M_{\mu\nu}, \tag{2.40}$$

where Λ, the enigmatic cosmological parameter, is a spacetime constant. It is important to note that, analogous to the general relativistic case, this is the *only* admissible generalization of the field equations compatible with the hypotheses that (1) gravitation is a manifestation of spacetime geometry and (2) Newtonian mechanics is valid in the absence of gravitation [10]. Not surprisingly, however, it is possible to relax the restriction of spacetime constancy on Λ if the matter conservation condition (2.39) is concurrently relaxed.

2.2.3 An additional constraint on the curvature tensor

As far as the Newton-Cartan theory is viewed as a limiting form of Einstein's theory of gravity in which special relativistic effects become negligible, equations (2.21), (2.27), and (2.40) constitute the complete set of gravitational field equations of the theory [9]. The logical structure of the covariant Newtonian theory, however, is flexible enough [10] to allow an additional constraint on the curvature tensor, namely

$$h^{\lambda\sigma} R^{\alpha}{}_{\sigma\gamma\delta} \equiv R^{\alpha\lambda}{}_{\gamma\delta} = 0 \qquad (2.41)$$

(or, equivalently, $t_{[\sigma} R^{\alpha}{}_{\beta]\gamma\delta} = 0$ [6]), implying that parallel-transport of space-like vectors around a small closed curve is path-independent. Since this restriction, in physical terms, implies that "the rotation axes of freely falling, neighboring gyroscopes do not exhibit relative rotations in the course of time" (no *relative* rotations for timelike geodesics), in essence it just asserts a standard of rotation: the existence of 'absolute rotation' *à la* Newton (*"Gesetz der Existenz absoluter Rotation"* [12]). In fact, Newton-Cartan theory commonly discussed in the literature with its usual set of field equations (2.21), (2.27), and (2.40) is, strictly speaking, a slight generalization of Newton's original theory of gravity (although, *it* constitutes the true Galilean-relativistic limit of Einstein's theory in general, and *not* the classical Newtonian theory of gravity [10–13]). Put differently, unlike the usual Newton-Cartan field equations the condition (2.41) is *not* a necessary consequence of the Galilean-relativistic limit ('c $\to \infty$') of Einstein's field equations, but, instead, is an added restriction on the Newton-Cartan structure [12, 13]. Indeed, it is not possible to recover the Poisson equation $\Delta \Phi = 4\pi G \rho$ of Newton's theory from the usual Newton-Cartan field equations (2.21), (2.27), and (2.36) without any global assumptions unless this extra condition prohibiting any rotational holonomy is imposed on the curvature tensor [12]. It becomes redundant, however, if non-intersecting spacelike hypersurfaces covering the Galilean spacetime are required to asymptotically resemble Euclidean space [9, 11]. Such a global boundary condition of asymptotically flat spacetime — which idealizes the gravitating systems as isolated systems — is, of course, of great historical and physical importance not only in the case of Newton's theory, but also in the case of Einstein's theory. Nevertheless, any such global condition can only encompass special physical situations ('island universes'), and, in general, the extra condition (2.41) is inevitable to ensure smooth recovery of the Newtonian Poisson equation from the Newton-Cartan field equations. Therefore, in what follows, we shall view equation (2.41) as an extraneously imposed but necessary field equation on the Newton-Cartan structure.

For reasons that will become increasingly clear as we go on, it is convenient to express the above intrinsic condition (2.41) in terms of local variables u^ν and A_μ in analogy with the equation (2.29) representing the condition (2.27). The desired expression — which must hold at least locally — is

$$\nabla_{[\gamma} \nabla_{\delta]} A_\mu - \overset{u}{h}_{\mu\nu} \nabla_{[\gamma} \nabla_{\delta]} u^\nu = 0. \qquad (2.42)$$

Contracting with $h^{\alpha\mu}$, replacing $-v^\alpha$ for $h^{\alpha\mu} A_\mu - u^\alpha$, multiplying by $-\frac{2}{A^2} h^{\lambda\beta} A_\beta$, and using equation (2.34) reveals that this expression is equivalent to the intrinsic form (2.41). It tells us that, as long as the additional condition (2.41) on the curvature tensor is satisfied, path-independence of parallel-transport around a small closed curve holds true not only for the fiducial observer u (cf. equation (2.34)), but also for any observer v in 4-rotation with u.

Finally, we end this subsection by observing the logical completeness of the Newton-Cartan structure described in the last two subsections. First of all, straightforward evaluations from equation (2.26) show that the curvature tensor of the connection (2.24) satisfies the relations (2.27) and (2.41). Conversely, and more importantly for our purposes, Trautman [6] has shown that equations (2.27) and (2.41) together imply the existence of a scalar potential $\overset{v}{\Phi}$ and a flat connection $\overset{v}{\Gamma}$ such that the components of the 'curved' connection Γ satisfy the relation (2.24) and its curvature tensor satisfies equation (2.26). This necessary and sufficient equivalence of equation (2.26) with the pair (2.27) and (2.41) consistently rounds off the geometrization of Newton's theory of gravity due to Cartan. For convenience, let us rewrite the complete geometric set of gravitational field-equations of the Newton-Cartan theory:

$$h^{\alpha\beta}t_\beta = 0, \quad \nabla_\alpha h^{\beta\gamma} = 0, \quad \nabla_\alpha t_\beta = 0, \quad \partial_{[\alpha} t_{\beta]} = 0, \qquad (2.43.a)$$

$$R^\alpha{}_\beta{}^\gamma{}_\delta = R^\gamma{}_\delta{}^\alpha{}_\beta, \qquad (2.43.b)$$

$$R^{\alpha\lambda}{}_{\gamma\delta} = 0, \qquad (2.43.c)$$

$$\text{and} \quad R_{\mu\nu} + \Lambda t_{\mu\nu} = 4\pi G\, M_{\mu\nu}, \qquad (2.43.d)$$

$$(2.43)$$

where the first four equations specify the degenerate 'metric' structure and a set of torsion-free connections on $\mathcal{M}$, the fifth one picks out the Newton-Cartan connection from this set of generic possibilities, the sixth one postulates the existence of absolute rotation, and the last one relates spacetime geometry to matter in analogy with Einstein's field equation.

2.2.4 Gauge and Lie-algebraic structures of Newton- Cartan spacetime

2.2.4.1 The gauge structure

Since the 'æther-frame' with its four-velocity u has been adapted as an auxiliary structure into the description of Newton-Cartan framework delineated in the previous subsections, any physical theory constructed in accordance with this framework must be invariant under changes in u — and, hence, under associated changes in the 'vector-potential' A — if the theory is to maintain general covariance. Using a particular pair of potentials (u, A) rather than a given Newton-Cartan connection Γ amounts to choosing a 'Bargmann gauge' in the fiber-bundle formulation of the Newtonian gravity studied by Duval and Künzle [34]. They take Bargmann bundle $B(\mathcal{M})$ to be a principal bundle over the Galilean manifold $\mathcal{M}$ with the Bargmann group B (a non-trivial central extension of the inhomogeneous Galilean group $\mathcal{G}$ by an abelian one-dimensional phase group U(1) appearing in the exact sequence $1 \to \mathrm{U}(1) \to B \to \mathcal{G} \to 1$) as its structure group. It is a U(1)-extension of a sub-bundle $\mathcal{G}(\mathcal{M})$ of the bundle of Galilean-relativistic affine frames of $\mathcal{M}$, with a surjective principal bundle homomorphism $B(\mathcal{M}) \to \mathcal{G}(\mathcal{M})$, constructed as follows. Let $Gl(\mathcal{M})$ be the principal bundle of linear frames over $\mathcal{M}$. A *Galilean structure* is then a reduction of $Gl(\mathcal{M})$ to the homogeneous subgroup $\mathcal{G}_0 := SO(3) ⓈR^3$ of the Galilean group (where Ⓢ denotes a semidirect product), and the sub-bundle $\mathcal{G}(\mathcal{M})$ of the bundle of Galilean-relativistic affine frames of $\mathcal{M}$ is the pull-back of $\mathcal{G}_0(\mathcal{M})$ by the canonical projection $T\mathcal{M} \to \mathcal{M}$, where $T\mathcal{M}$ is the tangent bundle over $\mathcal{M}$. If $i_o : \mathcal{G}_0(\mathcal{M}) \hookrightarrow \mathcal{G}(\mathcal{M})$ denotes an embedding through the zero section of $T\mathcal{M}$, and $B(\mathcal{M})$ is a U(1)-extension of $\mathcal{G}(\mathcal{M})$ with structure group B, then let $B_0(\mathcal{M})$ be the pull back of $B(\mathcal{M})$ by i_o. The quotient bundle $P(\mathcal{M}) := B_0(\mathcal{M})/\mathcal{G}_0$ is then a U(1)-principal bundle over $\mathcal{M}$. Now, it can be shown [34] that any compatible Newton-Cartan connection defined by equation (2.27) above designates an entire class of connections on $P(\mathcal{M})$ which are in one-to-one correspondence with the sections u of the unit tangent bundle $T_1\mathcal{M} = \mathcal{G}_0(\mathcal{M})/SO(3)$ of the structure $(\mathcal{M}; \tau, h)$. Clearly, the sections u are nothing but the four-velocity observer fields considered above with their gauge-dependent spatial projection fields $\overset{u}{h}_{\mu\nu}$. The gauge theory of Newtonian gravity we are considering must, therefore, be covariant under changes of the sections u, over and above the covariance under the automorphisms of $P(\mathcal{M})$. Duval and Künzle show that the Newton-Cartan connection defined by the constraint (2.27) on the Galilean structure (2.21) is indeed invariant under the simultaneous changes

$$\chi \mapsto \tilde{\chi} = \chi + f \qquad (2.44a)$$
$$u^\alpha \mapsto \tilde{u}^\alpha = u^\alpha + h^{\alpha\sigma}\mathrm{w}_\sigma \qquad (2.44b)$$
$$A_\alpha \mapsto \tilde{A}_\alpha = A_\alpha + \partial_\alpha f + \mathrm{w}_\alpha - (u^\sigma \mathrm{w}_\sigma + \tfrac{1}{2}h^{\mu\nu}\mathrm{w}_\mu \mathrm{w}_\nu)t_\alpha \qquad (2.44c)$$
$$(2.44)$$

where χ is the U(1)-phase (treated as the only intrinsic group coordinate), $f \in C^\infty(\mathcal{M}, \mathbb{R})$ is an arbitrary smooth map $\mathcal{M} \to \mathbb{R}$, w is a Galilean boost 1-form (defined only modulo t_α) belonging to the space of covector-fields $\Omega^1(\mathcal{M})$ on $\mathcal{M}$, and A_α is the 'vector-potential' defined above. The transformations (2.44a) and (2.44b) can be regarded as the 'vertical automorphisms' of the unit tangent bundle $B_0(\mathcal{M})/SO(3)$, and along with the diffeomorphisms $\phi \in \mathrm{Diff}(\mathcal{M})$ of $\mathcal{M}$ they compose the complete automorphism group

$$\mathcal{A}ut(B(\mathcal{M})) := \{(\phi, \mathrm{w}, f)| \phi \in \mathrm{Diff}(\mathcal{M}), \mathrm{w} \in \Omega^1(\mathcal{M}), f \in C^\infty(\mathcal{M}, \mathbb{R})\} \quad (2.45)$$

of the Newton-Cartan theory of gravity. Conversely, the projections of the fiber-preserving elements of the group $\mathcal{A}ut(B(\mathcal{M}))$ by the bundle projection map

$\pi : B(\mathcal{M}) \to \mathcal{M}$ are the smooth coordinate transformations of $\mathcal{M}$ into itself which constitute the diffeomorphism group $\text{Diff}(\mathcal{M})$; and, since the projection map π is a group homomorphism, the kernel of this map is the group

$$\mathcal{V}(B(\mathcal{M})) := [\Omega^1(\mathcal{M}) \times C^\infty(\mathcal{M}, \mathbb{R})] \qquad (2.46)$$

of vertical gauge transformations defined by equation (2.44). Thus, the group of bundle automorphism (2.45) encapsulates two classes of invariance which must be respected by the action $\mathcal{I}$ of any gauge-invariant field theory compatible with the Newton-Cartan structure $(\mathcal{M}; \tau, h, \Gamma)$. Mathematically, the elements of one of these two classes correspond to transforming the *fibers* of the bundle space $B(\mathcal{M})$ by the action of the normal (or invariant) subgroup $\mathcal{V}(B(\mathcal{M}))$ of the complete automorphism group $\mathcal{A}ut(B(\mathcal{M}))$, whereas the elements of the other class correspond to transforming the *base space* $\mathcal{M}$ of the bundle by the action of the factor subgroup $\text{Diff}(\mathcal{M}) = \mathcal{A}ut(B(\mathcal{M}))/\mathcal{V}(B(\mathcal{M}))$. This means that the automorphism group $\mathcal{A}ut(B(\mathcal{M}))$ has the structure of a semidirect product $\mathcal{V}(B(\mathcal{M})) \,\circledS\, \text{Diff}(\mathcal{M})$ indicating its status in the exact sequence

$$1 \longrightarrow \mathcal{V}(B(\mathcal{M})) \longrightarrow \mathcal{A}ut(B(\mathcal{M})) \longrightarrow \text{Diff}(\mathcal{M}) \longrightarrow 1 \,, \qquad (2.47)$$

and it acts on the set $\{(h, \tau, u, A)\}$ by

$$(\phi, \mathrm{w}, f) : \begin{pmatrix} h \\ \tau \\ u \\ A \end{pmatrix} \longmapsto \phi_* \begin{pmatrix} h \\ \tau \\ u + h(\mathrm{w}) \\ A + df + \mathrm{w} - \{\mathrm{w}(u) + \tfrac{1}{2}h(\mathrm{w}, \mathrm{w})\}\tau \end{pmatrix} \qquad (2.48)$$

giving the semidirect product group multiplication law:

$$(\phi_1, \mathrm{w}_1, f_1)(\phi_2, \mathrm{w}_2, f_2) = (\phi_1\phi_2, \, \phi_2^* \mathrm{w}_1 + \mathrm{w}_2, \, \phi_2^* f_1 + f_2) \,, \qquad (2.49)$$

34

where ϕ_* is the push-forward map and ϕ^* is the pull-back map corresponding to the diffeomorphism $\phi \in \mathrm{Diff}(\mathcal{M})$. More significantly, this complete gauge group $\mathcal{A}ut(B(\mathcal{M}))$ acts on the Newton-Cartan structure $(\mathcal{M}, h, \tau, \Gamma)$ only via the quotient subgroup $\mathrm{Diff}(\mathcal{M})$,

$$(\phi, \mathrm{w}, f) : \begin{pmatrix} h \\ \tau \\ \Gamma \end{pmatrix} \longmapsto \phi_* \begin{pmatrix} h \\ \tau \\ \Gamma \end{pmatrix}, \tag{2.50}$$

exhibiting that the principle of general covariance has been quite consistently adapted from Einstein's theory of gravity to this Newton-Cartan-Bargmann framework (cf. footnote 2).

2.2.4.2. The Lie-algebraic structure

The degenerate 'metric' structure of general Galilean spacetime permits some plasticity in the Lie-algebraic structure $T_e\,\mathcal{A}ut(B(\mathcal{M}))$ of the automorphism group $\mathcal{A}ut(B(\mathcal{M}))$ (here e is the unit element of the group). At least *six* different nested Lie algebras of infinitesimal 'isometries' (which play a role analogous to that of the algebra of Killing vector fields of Lorentzian spacetime) have been shown [33] to naturally arise as possible candidates for Newton-Cartan symmetry algebras. These nested Lie algebras include two extreme cases: (1) the *Coriolis algebra* — i.e., the Lie algebra of the infinite-dimensional Leibniz group [35] (the symmetry group of most general 'metric' automorphisms of the Galilean-relativistic spacetime) — and (2) the all important *Bargmann algebra* — i.e., the Lie algebra of the Bargmann group (the fundamental symmetry group of massive, non-interacting Galilean-relativistic systems — classical or quantal).

To see how these two Lie-algebraic structures arise as extreme cases, recall that Newton-Cartan connection Γ can be expressed in terms of the vector fields

u and v, and the scalar potential $\overset{v}{\Phi}$ (cf. equations (2.24), (2.7) and (2.32)). In terms of these variables the action (2.48) of the group $\mathcal{A}ut(B(\mathcal{M}))$ on the full Newton-Cartan-Bargmann structure $(\mathcal{M};\, h,\, \tau,\, u,\, v,\, \overset{v}{\Phi}\,)$ can be expressed as

$$
\begin{pmatrix} h \\ \tau \\ u \\ v \\ \overset{v}{\Phi} \end{pmatrix} \longmapsto \phi_* \begin{pmatrix} h \\ \tau \\ u + h(\mathrm{w}) \\ v + h(df) \\ \overset{v}{\Phi} + v(df) + \tfrac{1}{2} h(df,\, df) \end{pmatrix}, \tag{2.51}
$$

whereas the corresponding infinitesimal action of this automorphism group on the structure $(\mathcal{M};\, h,\, \tau,\, u,\, v,\, \overset{v}{\Phi}\,)$ can be seen as [33]

$$
\delta \begin{pmatrix} h \\ \tau \\ u \\ v \\ \overset{v}{\Phi} \end{pmatrix} = \begin{pmatrix} \pounds_{\mathrm{x}} h \\ \pounds_{\mathrm{x}} \tau \\ \pounds_{\mathrm{x}} u + h(\theta) \\ \pounds_{\mathrm{x}} v + h(dg) \\ \mathrm{x}(\overset{v}{\Phi}) + v(g) \end{pmatrix}, \tag{2.52}
$$

where $\theta \in \Omega^1(\mathcal{M})$, $\mathrm{g} \in C^\infty(\mathcal{M})$, and $\pounds_{\mathrm{x}}$ denotes a Lie derivative with respect to the symmetry generating vector-fields $\mathrm{x} = \mathrm{x}^\alpha \partial_\alpha$ on $\mathcal{M}$. The associated Lie-algebraic structure of this infinitesimal action works out to be

$$
[(\mathrm{x},\, \theta,\, \mathrm{g}),\, (\mathrm{x}',\, \theta',\, \mathrm{g}')] \;=\; ([\mathrm{x},\, \mathrm{x}'],\, \pounds_{\mathrm{x}} \theta' - \pounds_{\mathrm{x}'} \theta,\, \mathrm{x}(\mathrm{g}') - \mathrm{x}'(\mathrm{g})\,). \tag{2.53}
$$

If we now set $0 = \delta h = \delta \tau = \delta u = \delta v = \delta \overset{v}{\Phi}$ and, thereby, look for the stabilizer of the Newton-Cartan-Bargmann structure for the case of flat spacetime, we recover the Lie algebra of the Bargmann group [33]. Thus, the isotropy subgroup (or the stabilizer) of the full automorphism group $\mathcal{A}ut(B(\mathcal{M}))$ corresponding to the immutable flat structure ($h = \delta^{ab}\partial_a \otimes \partial_b$, $\tau = dt$, $u = \partial_0$, $A = 0$) is nothing but the Bargmann group, as one would expect. By relaxing one or more of the restrictions $0 = \delta u = \delta v = \delta\overset{v}{\Phi}$, and/or imposing various different restrictions on the connection Γ, a variety of intermediate algebraic structures associated with some physically interesting special cases may be worked out [33]. All of these intermediate symmetry groups are necessarily wider than the stabilizing Bargmann group. They form different subgroups of the Leibniz group — the most general infinite-dimensional 'isometry' group of the Newton-Cartan structure. The generators of Leibniz group are restricted only by the conditions $\mathcal{L}_x h = 0$ and $\mathcal{L}_x \tau = 0$, and, in general, do not Lie-transport the connection: $\mathcal{L}_x \Gamma \neq 0$. Consequently, the infinite-dimensional Lie algebra — the Coriolis algebra — corresponding to the Leibniz group preserves the absolute structure $(\mathcal{M}; h, \tau)$ of the Newton-Cartan spacetime, but leaves the connection Γ completely malleable, or dynamical. The elements of this most general symmetry group of the absolute structure represented in an arbitrary rigid frame take the form

$$
\begin{aligned}
x^0 &= x^0 + c^0 \equiv t + c^0 \,, \\
x^a &= O^a{}_b(t)\, x^b + c^a(t) \,, \qquad (a,b = 1,2,3),
\end{aligned}
\qquad (2.54)
$$

where $O^a{}_b(t) \in \mathrm{SO}(3)$ form an orthogonal rotation matrix for each value of t, $c^a(t) \in \mathbb{R}^3$ are arbitrary functions of $t \in \mathbb{R}$, and $c^0 \in \mathbb{R}$ is an infinitesimal time translation. Physically, this infinite-dimensional symmetry group correspond to transformations that connect different Galilean observers in arbitrary (accelerating and rotating) relative motion.

One physically and historically important subgroup of Leibniz group is the Milne group [36], which results as a direct consequence of the additional constraint (2.41) on the curvature tensor. In the above notation, it simply eliminates the time-dependence of the rotation matrix $O^a{}_b$ in equation (2.54), and may also be characterized by the restriction

$$\mathcal{L}_x(h^{\nu\sigma}\Gamma_\mu{}^\alpha{}_\sigma) \equiv \mathcal{L}_x\Gamma_\mu^{\alpha\nu} = 0 \qquad (2.55)$$

imposed directly on the connection [33] in addition to the conditions $\mathcal{L}_x h = 0$ and $\mathcal{L}_x \tau = 0$ on the metrics. The vector-fields x then constitute a Lie algebra corresponding to the infinitesimal Milne transformations

$$\begin{aligned} x^0 &= t + c^0\,, \\ x^a &= O^a{}_b\, x^b + c^a(t)\,, \qquad (a,b=1,2,3), \end{aligned} \qquad (2.56)$$

discussed in Ref. [29].

2.2.5 Derivation of the matter conservation laws

As is well-known [37] [28], in general relativity the principle of general covariance is sufficient for a derivation of the local differential conservation law $\nabla_\mu T^{\mu\nu} = 0$ for relativistic matter and non-gravitational fields from variation of an action functional with respect to the Lorentzian metric $g_{\mu\nu}$. The principle of general covariance in variational formulation of that theory is encapsulated in the mathematical requirement of invariance of the gravitation plus matter action $\mathcal{I}[g_{\mu\nu}, \Psi] := \mathcal{I}_g[g_{\mu\nu}] + \mathcal{I}_m[g_{\mu\nu}, \Psi]$ under the diffeomorphisms of the Lorentzian spacetime $(\mathcal{M}; g_{\mu\nu})$; i.e., in the requirement that if the map $^{(s)}\phi : \mathcal{M} \to \mathcal{M}$ defines a one-parameter family of diffeomorphisms, then $\mathcal{I}[^{(s)}\phi_* g_{\mu\nu}, {}^{(s)}\phi_* \Psi] = \mathcal{I}[g_{\mu\nu}, \Psi]$, where $^{(s)}\phi_*$ is the push-forward map corresponding to $^{(s)}\phi \in \mathrm{Diff}(\mathcal{M})$, and Ψ represents the matter fields. Moreover, general covariance demands that the matter action $\mathcal{I}_m$ by itself must be invariant under these diffeomorphisms if it were to retain an unequivocal physical meaning. In other words, for such diffeomorphic variations we must have

$$0 \;=\; \frac{d}{ds}\,\mathcal{I}_m[g_{\mu\nu},\,\Psi] \;\equiv\; \delta\mathcal{I}_m[g_{\mu\nu},\,\Psi] \;=\; \int \frac{\delta\mathcal{I}_m}{\delta g_{\mu\nu}}\,\delta g_{\mu\nu} \;+\; \int \frac{\delta\mathcal{I}_m}{\delta\Psi}\,\delta\Psi \;. \qquad (2.57)$$

If we now define

$$\frac{1}{2}\,T^{(\mu\nu)} \;=\; \frac{1}{2}\,T^{\mu\nu} \;:=\; \frac{1}{\sqrt{-g}}\,\frac{\delta\mathcal{I}_m}{\delta g_{\mu\nu}}\;, \qquad (2.58)$$

and assume that the matter fields satisfy the Euler-Lagrange equations $\frac{\delta\mathcal{I}_m}{\delta\Psi} = 0$, then equation (2.57) amounts to

$$0 \;=\; \delta\mathcal{I}_m \;=\; \int_{\mathcal{O}} \tfrac{1}{2}\,T^{\mu\nu}\delta g_{\mu\nu}\,dv$$
$$\qquad\qquad\qquad\qquad (2.59)$$

for some compact region $\mathcal{O} \subset \mathcal{M}$ with a non-null boundary $\partial\mathcal{O}$, where dv is the Lorentzian 4-volume element on $\mathcal{M}$. Using the well-known relation $\delta g_{\mu\nu} = \mathcal{L}_{X}\,g_{\mu\nu}$ $= 2\,\nabla_{(\mu}X_{\nu)}$ for such variations (where $\mathcal{L}_{X}$ denotes Lie derivative with respect to an arbitrarily chosen smooth vector field X_{μ} on $\mathcal{M}$) the above functional equation can be rewritten as

$$0 = \int_{\mathcal{O}} T^{\mu\nu}\nabla_{\mu}X_{\nu}\,dv = \int_{\mathcal{O}} \nabla_{\mu}(T^{\mu\nu}X_{\nu})\,dv - \int_{\mathcal{O}} (\nabla_{\mu}T^{\mu\nu})X_{\nu}\,dv = -\int_{\mathcal{O}} (\nabla_{\mu}T^{\mu\nu})X_{\nu}\,dv.$$
$$\qquad\qquad\qquad\qquad (2.60)$$

Here the last equality is obtained by converting the volume integral $\int_{\mathcal{O}} \nabla_{\mu}(T^{\mu\nu}X_{\nu})$ dv to a surface integral, which vanishes because X_{μ} vanishes on the boundary of the volume by assumption. Finally, since the region $\mathcal{O}$ and the vector field X_{μ} are arbitrarily chosen, the desired conservation law for the stress-energy of matter and non-gravitational field, $\nabla_{\mu}T^{\mu\nu} = 0$, immediately follows from this equation [28, 37].

Thus, in Lorentzian spacetime the stress-energy tensor $T^{\mu\nu}$ is seen to be dual to the metric $g_{\mu\nu}$ in the sense that variations of the matter action $\mathcal{I}_m$ with respect to the semi-Riemannian metric lead to the conserved matter tensor

$T^{\mu\nu}$. This state of affairs naturally suggests that in our Galilean-relativistic spacetime there should be *a pair* of quantities $(S_{\mu\nu}, C^\mu)$ dual to the pair of metrics $(h^{\mu\nu}, t_\mu)$ playing a role analogous to that of the relativistic stress-energy tensor $T^{\mu\nu}$. However, as shown by Künzle and Duval [11, 30], if we consider an action functional analogous to (2.59) defined on the Galilean spacetime $\mathcal{M}$, say

$$\delta\mathcal{I}_m = \int_{\mathcal{O}} \{\tfrac{1}{2}S_{\mu\nu}\delta h^{\mu\nu} + C^\mu\delta t_\mu\}\, \wp\, d^4x, \qquad (2.61)$$

with $\wp\, d^4x$ being the Galilean 4-volume element (cf. equation (2.3)), then the equations following from it do not correctly correspond to the well-known balance equations of energy and momentum (they lack an essential acceleration term of the classical equations.) The culprit, of course, is the degenerate or non-semi-Riemannian structure of the Galilean spacetime; as we have often noted, the degenerate pair of metrics $(h^{\mu\nu}, t_\mu)$ does not determine the connection completely. Fortunately, the difficulty also suggests its resolution. As we now know, introduction of the pair (u^α, A_α) of gauge fields as an auxiliary structure in the Galilean spacetime does fix the connection uniquely, and, hence, all we need to do is to enlarge the set of variables in the argument of the above action functional by these two gauge variables. But then the principle of general covariance demands that the variation of the resulting action, $\mathcal{I}_m[h^{\mu\nu}, t_\nu, A_\mu, u^\nu, \Psi]$, must be invariant not just under the diffeomorphisms of $\mathcal{M}$, but also under the vertical transformations (2.44) of these gauge variables; i.e., the variations of the matter action must be invariant under the entire gauge group $\mathcal{A}ut(B(\mathcal{M}))$ of the Newton-Cartan theory discussed in the previous subsection. Accordingly, following Duval and Künzle [30, 34], we augment (2.61) to be the functional

$$\delta\mathcal{I}_m = \int_{\mathcal{O}} \{\tfrac{1}{2}S_{\mu\nu}\delta h^{\mu\nu} + C^\mu\delta t_\mu + J^\mu\delta A_\mu + K_\mu\delta u^\mu\}\, \wp\, d^4x \qquad (2.62)$$

of four different variations, $\delta h^{\mu\nu}$, δt_μ, δA_μ, and δu^μ, and require it to be invariant under the full gauge group $\mathcal{A}ut(B(\mathcal{M}))$. Then, not only the equations of motion for matter fields derived from the variations of this action are gauge-invariant under $\mathcal{A}ut(B(\mathcal{M}))$ (see the following section for an example), but the associated matter-current density J^μ and 'Hilbert' stress-energy tensor[4]

$$N^\mu{}_\nu := h^{\mu\sigma} S_{\sigma\nu} - C^\mu t_\nu + J^\mu(v_\nu - \tfrac{1}{2}v^2 t_\nu) + u^\mu K_\nu, \qquad (2.63)$$

are also invariant under the action of that group, where $\rho := J^\sigma t_\sigma$, $J^\sigma := \rho v^\sigma$, $v_\sigma := \overset{u}{h}_{\sigma\alpha} v^\alpha$, and $v^2 := v^\sigma v_\sigma$.

(*4. Following Duval and Künzle we call this a 'Hilbert' stress-energy tensor because it corresponds to the general relativistic matter tensor (2.58) obtained by variations with respect to the Lorentzian metric. The canonical stress-energy tensor implied by Noether's theorem is different and not gauge-invariant.*)

What is more, the matter-current density and the 'Hilbert' stress-energy tensor satisfy the balance equations

$$\nabla_\mu J^\mu = 0 \qquad\qquad (2.64a)$$

and $\qquad \nabla_\mu N^\mu{}_\nu = \rho \, \overset{v}{h}_{\nu\sigma} \, v^\alpha \nabla_\alpha v^\sigma \qquad\qquad (2.64b)$

$$(2.64)$$

(as meticulously shown by Duval and Künzle [30]), which now correctly reduce to the classical expressions on flat spacetime [11]. Thus, the relativistic stress-energy tensor decouples into the pair $(N^\mu{}_\nu, J^\mu)$ in this nonrelativistic theory, describing the stress and energy flow by the tensor $N^\mu{}_\nu$ and the matter flow by the vector J^μ. This distinct role of the concept of mass-current from that of the stress-energy flow in the Newton-Cartan theory is closely related to the well-known fact that mass plays distinctly different roles in the Galilean and Poincaré invariant mechanics.

If we now define a mass-momentum-stress tensor (or matter tensor, for short) by

$$M^{\mu\nu} := -h^{\nu\sigma} N^\mu{}_\sigma + J^\mu v^\nu, \qquad (2.65)$$

41

then the balance equation (2.64b) immediately yields the nonrelativistic matter conservation law

$$\nabla_\mu M^{\mu\nu} \;=\; 0 \qquad\qquad (2.66)$$

analogous to the relativistic conservation law $\nabla_\mu T^{\mu\nu} = 0$. As a matter of fact, (2.66) is precisely the Galilean-relativistic limit-form of the relativistic conservation law, as shown by Künzle [11]. Substituting the expression (2.63) into the definition (2.65) of the matter tensor, and using the relation $K_\nu = -\overset{u}{h}_{\nu\sigma} J^\sigma + (constant)\, t_\nu$ (which has been derived by Duval and Künzle in [30]), we obtain the tensor $M^{\mu\nu}$ in a more transparent form:

$$M^{\mu\nu} \;=\; \rho\, u^\mu u^\nu \,-\, 2\rho\, u^{(\mu}h^{\nu)\sigma}A_\sigma \,-\, h^{\mu\sigma}h^{\nu\alpha}S_{\sigma\alpha}\,, \qquad (2.67)$$

which is identical to the mass-momentum-stress tensor (2.38) if we identify $s^\nu \equiv \rho\, h^{\nu\sigma}A_\sigma$. Consequently, the matter conservation law (2.66) is identical to the condition (2.39), which, at that stage, we had to impose independently. Amiably enough, in the present variational approach it is a *derived* result, following directly from the principle of general covariance, provided, of course, the matter field equations $\frac{\delta \mathcal{I}_m}{\delta \Psi} = 0$ are satisfied.

2.3 Quantum field theory on the curved Newton -Cartan spacetime

2.3.1 One-particle Schrödinger theory

In the previous section we reviewed the classical Newton-Cartan theory in some detail. The first systematic study of the *quantum* theory of freely falling particles in (unquantized) Newton-Cartan spacetime with illustrations of how the principle of equivalence works for such quantum systems has been carried out by Kuchař [1]. He arrives at a generally-covariant version of the Schrödinger's equation for a free quantum particle in an arbitrary, noninertial frame in such a classical spacetime which, in the special case of the Galilean frame with its distinguished gravitational potential, reduces to the usual Galilean-relativistic Schrödinger equation. Following the 'parameterized' canonical formalism [38], Kuchař starts with an action integral, giving Newton-Cartan geodesics as the extremal paths, as it appears to an arbitrary observer in an arbitrary gauge. After 'deparameterizing' it by labelling the worldlines with the Newtonian absolute time t, he casts the action into the generalized Hamiltonian form to prepare for the Dirac's constraint quantization. The transformation to the quantum theory is then easily achieved in an unambiguous, coordinate independent fashion by quantizing the motion of the particle with the use of the Dirac method. The resulting quantum mechanical equation of motion, or the quantum mechanical analog of the geodesic equation, turns out to be nothing but a covariant version of the ordinary Schrödinger equation for the free particle, with gravitational forces absorbed in the structure of spacetime in which the particle is freely falling. Thus, in a nutshell, Kuchař has successfully shown that quantum mechanics remains consistent in the presence of the Newtonian connection-field viewed as an effect of the curving of spacetime, and, due to the unique foliation possessed by the Newton-Cartan spacetime, Galilean-relativistic quantum theory escapes 'the problem of time' [3, 39] usually encountered in the attempts to canonically quantize parameterized dynamical systems in the presence of Einsteinian connection-field. In other words, Kuchař has shown that it is possible to unequivocally quantize Galilean-relativistic parameterized systems by means of the Dirac method such that the evolution of their quantum state $\Psi : \mathcal{M} \to \mathbb{C}$, unlike their general-relativistic counterpart, does not depend on the choice of spacetime foliation (which, of course, is uniquely given in the Newtonian case). Consequently, the corresponding Hilbert-space inner-product $\langle \Psi_1 \Psi_2 \rangle_t$ remains the same for all such domains of simultaneity. It is worth noting here that in a subsequent work De Bièvre [40] has obtained essentially the same covariant Schrödinger-Kuchař equation (in the framework of the Bargmann bundle discussed in the previous section) directly from an application of the principle of equivalence rather than *a posteriori* exhibiting the compatibility of this principle with the correct quantum dynamics of a free test particle in a gravitational field *à la* Kuchař. This further justifies the validity of the principle of equivalence in the quantum domain, and, to put more strongly, logically *demands* the generally-covariant reformulation of the Galilean-relativistic quantum dynamics.

For our purposes in this paper, however, it is convenient to follow the covariant framework of Duval and Künzle [34], who show that a prescription of minimal (Newtonian) gravitational coupling naturally leads to the four-dimensional spacetime-covariant Schrödinger-Kuchař equation as a result of extremization of an action of the form (2.62) discussed in the previous section. Recall the U(1)-principle bundle $\mathcal{M}'$ from the subsection 2.2.4, and consider a vector bundle E with the Hilbert space $L^2(\mathbb{R}^3)$ of square integrable functions on $\mathbb{R}^3$ as the standard fiber associated with it. Under the gauge transformation (2.48) a section Ψ of E, or a wave-function, changes as

$$(\phi, \mathrm{w}, f) : \Psi \longmapsto \phi_*[\exp(i\frac{m}{\hbar}f)\Psi] \; ; \tag{2.68}$$

and, in analogy with the electromagnetic gauge theory, the covariant derivative induced by the connection on $\mathcal{M}'$ acting on sections of E is

$$\mathrm{D}_\alpha \Psi := (\partial_\alpha - i\frac{m}{\hbar}A_\alpha)\Psi \; . \tag{2.69}$$

Now, a Lagrangian density for the free one-particle Schrödinger equation

$$\frac{\hbar^2}{2m}\delta^{ab}\partial_a\partial_b\Psi + i\hbar\,\partial_0\Psi = 0 \qquad (a,b=1,2,3) \tag{2.70}$$

may be taken to be

$$\mathcal{L}_{Schr} = \frac{\hbar^2}{2m}\delta^{ab}\partial_a\Psi\partial_b\overline{\Psi} + i\frac{\hbar}{2}(\Psi\partial_0\overline{\Psi} - \overline{\Psi}\partial_0\Psi)\,, \tag{2.71}$$

which on the curved Newton-Cartan spacetime becomes

$$\mathcal{L}_{Schr} \longrightarrow \mathcal{L}_{Kuch} = \wp\,4\pi G\,\{\frac{\hbar^2}{2m}h^{\alpha\beta}D_\alpha\Psi\overline{D_\beta\Psi} + i\frac{\hbar}{2}u^\alpha(\Psi\overline{D_\alpha\Psi} - \overline{\Psi}D_\alpha\Psi) \tag{2.72}$$

if we use the contravariant 3-metric $h^{\alpha\beta}$ for the Laplacian in place of the Kronecker delta δ^{ab}, write $\wp\,d^4x$ (cf. Eq. (2.3)) for the spacetime volume element on $\mathcal{M}$, and replace $\partial_0\Psi$ by $u^\alpha D_\alpha\Psi$ (the propagation covariant derivative along the timelike vector field u^α) and $\partial_\alpha\Psi$ by $D_\alpha\Psi$ in accordance with the minimal interaction principle (the meaning of the multiplicative factor $4\pi G$ will become clear in the following section). This latter Lagrangian density is manifestly covariant with respect to the diffeomorphisms $\phi \in \mathrm{Diff}(\mathcal{M})$ of $\mathcal{M}$ since $h^{\alpha\beta}$, t_α, A_α, u^α, Ψ, and $\overline{\Psi}$ are all tensor fields on the spacetime. What is more, it is also invariant under the simultaneous vertical gauge transformations (2.44) of A, u, Ψ, and $\overline{\Psi}$. Consequently, we expect the resulting Schrödinger-Kuchař theory to be independent of the choice of a Galilean observer represented by u with the corresponding gravitational interaction being uniquely described by the Newton-Cartan structure $(\mathcal{M}; h, \tau, \nabla)$ alone. The Euler-Lagrange equations corresponding to the Lagrangian density (2.72) obtained by *independently* extremizing the corresponding action functional $\mathcal{I}_{Kuch}$ with respect to $\overline{\Psi}$ and Ψ are

$$\mathcal{E}_{Kuch}[\Psi] = [\frac{\hbar^2}{2m}D^\alpha D_\alpha + i\hbar u^\alpha D_\alpha + i\frac{\hbar}{2}\nabla_\alpha u^\alpha]\Psi = 0 \tag{2.73}$$

and its complex conjugate, respectively; and once the gauge-covariant derivatives D_α in equation (2.73) are worked out, we indeed obtain the desired covariant Schrödinger-Kuchař equation on the curved Newton-Cartan spacetime:

$$[\frac{\hbar^2}{2m}\nabla^\alpha\partial_\alpha + i\hbar(u^\alpha - h^{\alpha\beta}A_\beta)\partial_\alpha + m(u^\alpha A_\alpha - \tfrac{1}{2}h^{\alpha\beta}A_\alpha A_\beta) + i\frac{\hbar}{2}\nabla_\alpha(u^\alpha - h^{\alpha\beta}A_\beta)]\Psi = 0.$$
$$(2.74)$$

Among many interesting results in this theory [1, 34], the two we will need later are the expressions for the conserved matter-current density (cf. Eq. (2.64a))

$$J^\alpha := \frac{1}{\wp\,4\pi G}\frac{\delta\mathcal{I}_{Kuch}}{\delta A_\alpha} = m\Psi\overline{\Psi}\,u^\alpha + i\frac{\hbar}{2}\,h^{\alpha\beta}(\Psi\overline{D_\beta\Psi} - \overline{\Psi}D_\beta\Psi)\,, \quad \nabla_\alpha J^\alpha = 0\,,$$
$$(2.75)$$

where $\rho \equiv m\Psi\overline{\Psi}$ is the invariant mass density, and, more significantly, the conserved matter tensor (cf. Eq. 2.67)

$$M^{\mu\nu} = m\Psi\overline{\Psi}\,u^\mu u^\nu - 2\,m\Psi\overline{\Psi}\,u^{(\mu}h^{\nu)\sigma}A_\sigma - S^{\mu\nu}\,, \quad \nabla_\mu M^{\mu\nu} = 0\,, \quad (2.76)$$

with

$$S^{\mu\nu} = \{\tfrac{1}{2}h^{\sigma\alpha}\Omega_{\alpha\sigma} + u^\sigma\Omega_\sigma - m\Psi\overline{\Psi}\,u^\sigma A_\sigma\}h^{\mu\nu} - 2\,u^{(\mu}h^{\nu)\sigma}\Omega_\sigma - h^{\mu\sigma}h^{\nu\alpha}\Omega_{\sigma\alpha}\,, \quad (2.77)$$

$\Omega_{\mu\nu} := \frac{\hbar^2}{m}D_{(\mu}\Psi\,\overline{D_{\nu)}\Psi}$, and $\Omega_\mu := \frac{i\hbar}{2}(\Psi\partial_\mu\overline{\Psi} - \overline{\Psi}\partial_\mu\Psi)$. As the above covariant Schrödinger-Kuchař equation, both of these conserved flows appear here as quantities *derived* from the action functional, and their respective expressions are invariant under the combined diffeomorphisms and gauge transformations (2.48) as expected.

2.3.2 Galilean-relativistic quantum field theory

In the previous subsection we have reviewed one-particle Schrödinger mechanics on the curved Newton-Cartan spacetime. It is well-known [41–43], however, that such a mechanics on the usual flat Galilean spacetime can be interpreted also as a second-quantized field theory in which the wave-function Ψ becomes an operator-valued distribution in the Fock space of unspecified number of identical particles. Thanks to the above described minimal coupling formulation of the Schrödinger-Kuchař theory, it turns out that this Fock-space representation of many-particle system can be quite straightforwardly generalized to our case of curved background. Given the Lagrangian density (2.72), the first step towards constructing the alleged generalization is to evaluate the momenta conjugate to the classical c-number fields Ψ and $\overline{\Psi}$:

$$\overline{P} := \frac{\delta \mathcal{L}_{Kuch}}{\delta (u^\sigma D_\sigma \Psi)} = -2i\hbar \wp \pi G \overline{\Psi}, \tag{2.78}$$

and its complex conjugate. This relation suggests that we should not regard $\overline{\Psi}$ as an independent variable, but rather as proportional to the canonical conjugate of Ψ. Therefore, we introduce a new, more appropriate set of canonical variables $\{\psi, p\}$,

$$
\begin{aligned}
\psi &:= \frac{1}{\hbar\sqrt{8\pi G \wp}} \left(2\pi G \wp \hbar \overline{\Psi} + i\overline{P} \right), \\
p &:= \frac{1}{\sqrt{8\pi G \wp}} \left(P + i2\pi G \wp \hbar \Psi \right),
\end{aligned}
\tag{2.79}
$$

so that equation (2.78) and its conjugate yield

$$\psi = \sqrt{2\pi G \wp}\, \overline{\Psi} \qquad \text{and} \qquad p = i\hbar \overline{\psi}. \tag{2.80}$$

Given a spacelike hypersurface Σ_t (cf. subsection 2.4.2 below), the quantum theory can now be easily defined by the usual equal-time commutation $(-)$ or anticommutation $(+)$ relations,

$$
\begin{aligned}
[\hat{\psi}(x), \hat{\psi}(x')]_\mp &= 0\,, \quad [\hat{\psi}^\dagger(x), \hat{\psi}^\dagger(x')]_\mp = 0\,, \\
\text{and} \quad [\hat{\psi}(x), \hat{\psi}^\dagger(x')]_\mp &= \hat{\mathbb{1}}\, \delta(\vec{x} - \vec{x}')\,,
\end{aligned}
\tag{2.81}
$$

corresponding to the bosonic $(-)$ or fermionic $(+)$ algebra, where $\vec{x} \in \Sigma_t$, $\hat{\mathbb{1}}$ is the identity operator in the Fock space, and, as always, we have replaced the complex conjugate function $\overline{\psi}$ with the Hermitian conjugate operator $\widehat{\psi}^\dagger$. It is worth recalling the well-known fact [42] that *a Galilean-relativistic theory provides no connection between spin and statistics*. Put differently, the Schrödinger field can be quantized equally well by imposing either commutation or anticommutation relations on the field operators as we have done. In either case, the operator $\widehat{\psi}^\dagger$ acts as a creation operator, whereas its Hermitian conjugate $\widehat{\psi}$ acts as an annihilation operator, which is assumed to annihilate the Milne-invariant (cf. Eq. (2.56)) vacuum state:

$$\widehat{\psi}\,|0\rangle \;=\; 0\,. \tag{2.82}$$

All the operators needed to describe the Schrödinger field theory on the curved Newton-Cartan spacetime can now be constructed from these two operators. One of the most important among these is the number density operator $\widehat{\varrho} = \widehat{\psi}^\dagger\widehat{\psi}$ which, as a consequence of the condition (2.82), annihilates the vacuum: $\widehat{\varrho}\,|\,0\,\rangle = 0$. A related operator relevant to us here is the mass operator

$$\widehat{\mathbb{M}} \;:=\; m\int_{\Sigma_t} \widehat{\psi}^\dagger\widehat{\psi}\ d^3x\,, \tag{2.83}$$

which implies

$$[\widehat{\mathbb{M}},\,\widehat{\psi}^\dagger(x)]_\mp \;=\; m\,\widehat{\psi}^\dagger(x) \qquad \text{and} \qquad [\widehat{\mathbb{M}},\,\widehat{\psi}(x)]_\mp \;=\; -\,m\,\widehat{\psi}(x)\,. \tag{2.84}$$

As a result, we may also define a covariant mass-density operator

$$\widehat{M}_{\mu\nu} \;:=\; m\,\widehat{\psi}^\dagger\widehat{\psi}\,t_{\mu\nu}\,, \tag{2.85}$$

which will be discussed further in the last section. Another important operator of interest to us is the angular-momentum operator

$$\widehat{\mathbf{J}} \;:=\; \int_{\Sigma_t} \widehat{\psi}^\dagger\,\mathbf{s}\,\widehat{\psi}\ d^3x\,, \tag{2.86}$$

giving

$$[\hat{\mathbf{J}}, \hat{\psi}^\dagger(x)]_\mp = \mathbf{s}\,\hat{\psi}^\dagger(x) \quad \text{and} \quad [\hat{\mathbf{J}}, \hat{\psi}(x)]_\mp = -\mathbf{s}\,\hat{\psi}(x) \,. \tag{2.87}$$

In particular, a vanishing commutation with $\hat{\mathbf{J}}$ of a given field operator is inferred to imply that the field corresponds to spinless bosons.

As usual, the conjugate momentum (2.78) and its conjugate allows the construction of Hamiltonian density operator:

$$\begin{aligned}
\hat{\mathrm{H}}_{Kuch} &:= \hat{\mathrm{P}}^\dagger u^\sigma(\mathrm{D}_\sigma\hat{\Psi}) + u^\sigma(\mathrm{D}_\sigma\hat{\Psi})^\dagger \hat{\mathrm{P}} - \hat{\mathcal{L}}_{Kuch} \\
&= -4\pi G \wp \frac{\hbar^2}{2m} h^{\mu\nu}(\mathrm{D}_\mu\hat{\Psi})(\mathrm{D}_\nu\hat{\Psi})^\dagger \\
&= -\frac{\hbar^2}{2m} h^{\mu\nu}[\mathrm{D}_\mu(\sqrt{2}\,\hat{\psi})]^\dagger[\mathrm{D}_\nu(\sqrt{2}\,\hat{\psi})] \,. \tag{2.88}
\end{aligned}$$

Here the ordering of noncommuting operators is, of course, important, since, as it is obvious form equation (2.81), $\hat{\psi}$ and $\hat{\psi}^\dagger$ in this expression are not genuine operators but rather operator-valued distributions in the Fock space. We have made use of the usual normal-ordering prescription of keeping $\hat{\psi}^\dagger$ to the left of $\hat{\psi}$.

This completes our field-theoretic generalization of the Schrödinger theory on the curved Newton-Cartan spacetime. Of course, the equal-time (anti-)commutation relations (2.81) break manifest covariance of the theory. However, as we shall see in the subsection 2.4.3 below, this is not inevitable. What is physically more significant is the following observation. The theory we have constructed here is a quantum theory of *free* particles on the curved Newton-Cartan spacetime. If we now heuristically try to substitute the quantum operator (2.85) in the right-hand side of the field equation (2.40), we realize that what is needed for consistency is a theory in which these quantized Schrödinger particles *produce* the quantized Newton-Cartan connection-field through which they interact. In other words, consistency requires a generally-covariant quantum field theory of identical Schrödinger particles interacting through their own quantized Newtonian gravitational field. A construction of such an interacting quantum field theory is the goal of the remaining of this paper.

2.4 Newton-Cartan-Schrödinger theory from an action principle

Our aim in this section is, firstly, to obtain the complete Newton-Cartan-Schrödinger theory epitomized in equations (2.43), (2.64a), (2.66), and (2.74) from extremizations of a *single* $\mathcal{A}ut(B(\mathcal{M}))$-invariant action, and, then, to recast the theory into a *constraint-free* Hamiltonian form in 3+1-dimensions, as well as into a manifestly covariant reduced phase-space form in 4-dimensions, to pave the way for quantization in the next section.

2.4.1 Covariant Lagrangian formulation of the theory

Let us first segregate the dynamical variables from the non-dynamical variables. For this purpose, recall that the Newton-Cartan connection is dynamical in general and not an invariant backdrop: as discussed in the subsection 2.2.4 above, the most general group of symmetry transformations of the Galilean-relativistic structure — the Leibniz group — does not leave the connection invariant; the generators x of the Leibniz group forming an infinite-dimensional Lie algebra are constrained only by the conditions $\mathcal{L}_x t_{\mu\nu} = 0$ and $\mathcal{L}_x h^{\mu\nu} = 0$, and, in general, do not Lie-transport the Newton-Cartan connection, $\mathcal{L}_x \Gamma^{\gamma}{}_{\alpha\beta} \neq 0$. In other words, the respective 'isometries' $\mathcal{L}_x t_{\mu\nu} = 0$ and $\mathcal{L}_x h^{\mu\nu} = 0$ of the tensor fields $t_{\mu\nu}$ and $h^{\mu\nu}$ dictate that these fields are simply parts of the immutable background structure of the Newton-Cartan spacetime, and it is only the connection Γ of the full Newton-Cartan structure $(\mathcal{M}; h, \tau, \Gamma)$ which is left unrestrained by the Leibniz group[5].

(5. *Continuing our discussion in footnote 2 regarding the meaning of general covariance, it is worth emphasising that the metric fields $h^{\mu\nu}$ and $t_{\mu\nu}$ here defining (in Stachel's terminology [16]) the chronogeometrical structure of the Newton-Cartan gravity have been specified independently of the inertio-gravitational field $\Gamma_{\alpha}{}^{\gamma}{}_{\beta}$. Now in general relativity, due to the dynamical nature of the metric field $g_{\mu\nu}$, the meaninglessness of an a priori labelling of individual spacetime points is axiomatic: a point in the bare manifold $\mathcal{M}$ is not distinguishable form any other point — and, indeed, does not even become a point with physical meaning — until the metric field is dynamically determined. Unlike in general relativity, however, where the affine structure of spacetime is inseparably (and rather wholistically) identified with the inertio-gravitational field, in Newton-Cartan theory it is possible to individuate spacetime points as entia per se existing independently of, and logically prior to, the inertio-gravitational field $\Gamma_{\alpha}{}^{\gamma}{}_{\beta}$. This is because (again in Stachel's language) the metric fields $h^{\mu\nu}$ and $t_{\mu\nu}$ serve as non-dynamical individuating fields [16] specifying once and for all the immutable chronogeometrical structure, which in Newton-Cartan theory happens to be independent of the mutable inertio-gravitational structure. See [26], however, for a somewhat differing emphasis on the difficulties in pointwise identification of two different Newton-Cartan spacetimes.)*

Consequently, it is sufficient to treat the non-metric connection Γ as the *only* dynamical attribute determined by the matter distribution via the field equation (2.40). On the other hand, we recall that any Newton-Cartan connection can be affinely decomposed in terms of an æther-field u and an arbitrary vector-potential A as $\Gamma_{\alpha}{}^{\gamma}{}_{\beta} = \overset{u}{\Gamma}_{\alpha}{}^{\gamma}{}_{\beta} + \overset{A}{\Gamma}_{\alpha}{}^{\gamma}{}_{\beta}$ (cf. Eq. (2.30)). Therefore, it is sufficient to take the fields A and u of the full Newton-Cartan structure $(\mathcal{M}; h, \tau, u, A)$ as the only dynamical variables of the gravitational field when we proceed next to formulate the Newton-Cartan-Schrödinger theory in a Lagrangian form.

In what follows, however, it is convenient to view our theory as a *parameterized field theory* [38, 46] by introducing supplementary kinematical variables on which the two metrics depend. To see how this is done, consider an arbitrary covector-field $\overset{\circ}{A}_{\mu}$, an observer field $\overset{\circ}{u}^{\nu}$, the *classical* Schrödinger-Kuchař complex scalar field $\overset{\circ}{\Psi}$, and its complex conjugate $\overset{\circ}{\overline{\Psi}}$ as our dynamical field variables defined on a generic background structure $(\mathcal{M}; \overset{\circ}{h}, \overset{\circ}{\tau}, \overset{\circ}{\nabla})$ satisfying $\pounds_{\mathsf{x}} \overset{\circ}{t}_{\mu\nu} = 0 = \pounds_{\mathsf{x}} \overset{\circ}{h}^{\mu\nu}$ with an as-yet-unspecified connection $\overset{\circ}{\Gamma}$ given by equation (2.6) — i.e., let us not restrain the connection *a priori* to be Newton-Cartan by means of the constraint (2.27); in other words, at this juncture we do not impose the conditions (2.21) on this structure, and do not subject the arbitrary 1-form $\overset{\circ}{A} = \overset{\circ}{A}_{\mu}dx^{\mu}$ on $\overset{\circ}{\mathcal{M}}$ to satisfy the Newton-Cartan selection condition (2.28). Since the metrics $\overset{\circ}{h}^{\alpha\beta}$ and $\overset{\circ}{t}_{\beta}$ are fixed, an action functional defined on such a structure will not be invariant under all possible diffeomorphisms of $\mathcal{M}$. Therefore, we enlarge the configuration space of the theory by a new one-parameter family of kinematical variables $^{(s)}y \in \mathrm{Diff}(\mathcal{M})$ constituting the map

$$^{(s)}y \: : \: \mathcal{M} \longrightarrow \overset{\circ}{\mathcal{M}} \tag{2.89}$$

from a copy $\mathcal{M}$ of the manifold $\overset{\circ}{\mathcal{M}}$ to the manifold $\overset{\circ}{\mathcal{M}}$ itself, and think of $^{(s)}y(x)$ as a field on $\mathcal{M}$ taking values in $\overset{\circ}{\mathcal{M}}$. As is well-known, this is nothing but the procedure of *parameterization*; i.e., the procedure for obtaining a generally-covariant version of a field theory originally defined on some non-dynamical, background spacetime. Using the map (2.89), we can now pull-back the two metrics as well as the dynamical fields from the 'fixed' manifold $\overset{\circ}{\mathcal{M}}$ to the 'parameterized' manifold $\mathcal{M}$:

$$h^{\mu\nu}(y) \; := \; (y_{*})^{\mu}{}_{\alpha}(y_{*})^{\nu}{}_{\beta} \, \overset{\circ}{h}^{\alpha\beta} \, , \tag{2.90a}$$

$$t_\mu(y) := (y^*)^\alpha{}_\mu \mathring{t}_\alpha , \qquad\qquad\qquad (2.90b)$$

$$(2.90)$$

$A_\mu(y) := (y^*)^\alpha{}_\mu \mathring{A}_\alpha$, $u^\nu(y) := (y_*)^\nu{}_\mu \mathring{u}^\mu$, $\Psi(y) := y^* \mathring{\Psi}$, and $\overline{\Psi}(y) := y^* \overline{\mathring{\Psi}}$; where $(y^*)^\mu{}_\alpha$ denotes the induced map from the tangent space of $x \in \mathcal{M}$ to the tangent space of $y(x) \in \mathring{\mathcal{M}}$, or, equivalently, the pull-back map from the cotangent space of $y(x)$ to the cotangent space of x, and $(y_*)^\alpha{}_\mu$ denotes the inverse of $(y^*)^\mu{}_\alpha$.

Having defined the dynamical variables A_μ, u^ν, Ψ, and $\overline{\Psi}$ on the manifold $\mathcal{M}$ along with the supplementary kinematical variables ${}^{(s)}y$, our next concern is to construct a meaningful classical phase-space for the system using Lagrangian and Hamiltonian formalisms, and then define Poisson brackets on this space in order to proceed with quantization in the next section. Accordingly, we demand that extremizations of a stationary action defined on $\mathcal{M}$ with respect to variations of these variables lead to the complete set of gravitational field equations (2.21), (2.27), and (2.40), the matter conservation law (2.66), and the equation of motion (2.74) for the classical fields Ψ and $\overline{\Psi}$ on the curved spacetime, of the full Newton-Cartan-Schrödinger theory. A Lagrangian density which fulfils this demand can then be used to construct a Hamiltonian density, which would then lead us to the desired phase-space, and, subsequently, quantization of the theory can be accomplished by interpreting the functions on this phase-space as quantum mechanical operators.

Now, as discussed in the subsection 2.2.1, any Galilean spacetime $(\mathcal{M}; h, \tau, \nabla)$ is orientable since it possesses a canonical 4-volume element $\wp\, d^4x$ (cf. Eq. (2.3)), which is, conveniently, non-dynamical; in particular, $\nabla_\mu \varepsilon_{\alpha\beta\gamma\delta} = 0$. Unlike the general relativistic 4-volume element, which is determined from the dynamical gravitational field variables $g_{\mu\nu}$, here the volume element is derived using only the *non-dynamical* metric fields $h_{\mu\nu}$ and $t_{\mu\nu}$, and, more importantly, is independent of the dynamical field variables A_μ and u^ν. Furthermore, even in the absence of the conditions (2.21) defining the Galilean spacetime, one can begin with a measure form $\varepsilon_{\alpha\beta\gamma\delta}$ defined on the generic structure $(\mathcal{M}; h^{\mu\nu}, t_\nu)$ by equation (2.3). Therefore, if Z is the collection of all field configurations on the manifold $\mathcal{M}$, then we can form a tentative stationary action $\mathcal{I} : Z \to \mathbb{R}$ over some measurable region $\mathcal{O} \subset \mathcal{M}$ with a non-null boundary $\partial\mathcal{O}$ as follows:

$$\mathcal{I} = \int_{\mathcal{O}} d^4x\ \mathcal{L}(A_\nu, \nabla_\mu A_\nu, \nabla_\mu \nabla_\nu A_\alpha, u^\nu, \nabla_\mu u^\nu, \nabla_\mu \nabla_\nu u^\alpha, \Psi, \partial_\mu \Psi, \overline{\Psi}, \partial_\mu \overline{\Psi}; {}^{(s)}y)$$
(2.91)

where

$$\mathcal{L} \equiv \mathcal{L}_{Grav} + \mathcal{L}_{Kuch} \equiv [\mathcal{L}_{\mathrm{B}} + \mathcal{L}_\Gamma + \mathcal{L}_{\mathrm{N}} + \mathcal{L}_\Phi + \mathcal{L}_\Lambda], + [\mathcal{L}_\Psi + \mathcal{L}_{\mathrm{I}}], \quad (2.92a)$$

$$\mathcal{L}_{\mathrm{B}} \equiv +\wp\left\{\Upsilon_\mu h^{\mu\nu} t_\nu + \Upsilon^\mu_{\nu\sigma} \nabla_\mu h^{\nu\sigma} + \Upsilon^{\mu\nu} \nabla_\mu t_\nu + \tilde{\Upsilon}^{\mu\nu} \partial_{[\mu} t_{\nu]}\right\}, \quad (2.92b)$$

$$\mathcal{L}_\Gamma \equiv +\wp[\zeta\{u^\nu t_\nu - 1\} + \chi^{\mu\nu}\{\nabla_{[\mu} A_{\nu]} + \overset{u}{h}_{\sigma[\mu} \nabla_{\nu]} u^\sigma\}], \quad (2.92c)$$

$$\mathcal{L}_{\mathrm{N}} \equiv +\wp\,\chi^{\alpha\mu\nu}\{\nabla_{[\mu} \nabla_{\nu]} A_\alpha - \overset{u}{h}_{\alpha\sigma} \nabla_{[\mu} \nabla_{\nu]} u^\sigma\}, \quad (2.92d)$$

$$\mathcal{L}_\Phi \equiv +\wp\,\frac{\kappa}{2}\{h^{\mu\nu} \nabla_\mu \Theta \nabla_\nu \Theta\}, \qquad \Theta(A_\alpha, u^\sigma) \equiv \frac{1}{2} h^{\mu\nu} A_\mu A_\nu - A_\sigma u^\sigma, \quad (2.92e)$$

$$\mathcal{L}_\Lambda \equiv +\wp\,\chi\{t_\mu \nabla_\nu \chi^{\mu\nu} - (1-\kappa)\nabla_\sigma \Theta^\sigma + \lambda\,\Lambda_{\mathrm{N}} - \Lambda_{\mathrm{O}}\}, \qquad \Theta^\sigma := h^{\sigma\mu} \nabla_\mu \Theta, \quad (2.92f)$$

$$\mathcal{L}_\Psi \equiv +\wp\,4\pi G\left\{\frac{\hbar^2}{2m} h^{\alpha\beta} \partial_\alpha \Psi \partial_\beta \overline{\Psi} + i\frac{\hbar}{2} u^\alpha(\Psi \partial_\alpha \overline{\Psi} - \overline{\Psi} \partial_\alpha \Psi)\right\}, \quad (2.92g)$$

$$\mathcal{L}_{\mathrm{I}} \equiv -\wp\,4\pi G \int J^\alpha\, dA_\alpha, \quad (2.92h)$$

(2.92)

Υ_μ, $\Upsilon^\mu_{\nu\sigma}$, $\Upsilon^{\mu\nu}$, $\tilde{\Upsilon}^{\mu\nu} = \tilde{\Upsilon}^{[\mu\nu]}$, ζ, $\chi^{\mu\nu} = \chi^{[\mu\nu]}$, and $\chi^{\alpha\mu\nu} = \chi^{\alpha[\mu\nu]}$ are all (independent) undetermined Lagrange multiplier fields (symmetric in permutations of their indices unless indicated otherwise), κ and λ are *arbitrary* free parameters (with values *including* 0, 1, and, in case of λ, $\frac{\Lambda_{\mathrm{O}}}{\Lambda_{\mathrm{N}}}$), χ (also treated as a Lagrange multiplier field) will be seen in the next subsection to be the arbitrary function of gauge transformations (if $\chi \mapsto \chi + f$, then $A_\mu \mapsto A_\mu + \partial_\mu f$; cf. Eq. (2.44)), the scalars Λ_O and

$$\Lambda_N := t_\alpha \nabla_\sigma \nabla_\gamma \chi^{\alpha\sigma\gamma} \tag{2.93}$$

will turn out to be parts of the cosmological constant, and

$$J^\alpha \equiv m\Psi\overline{\Psi}\{u^\alpha - h^{\alpha\beta}A_\beta\} + i\frac{\hbar}{2}h^{\alpha\beta}\{\Psi\partial_\beta\overline{\Psi} - \overline{\Psi}\partial_\beta\Psi\}, \tag{2.94}$$

which is formally identical to the expression (2.75). Note that we have ensured the action $\mathcal{I}$ to have appropriate boundary terms by allowing the Lagrangian $\mathcal{L}$ to contain pure divergences. It is instructive to compare the Lagrangian density $\mathcal{L} \equiv \mathcal{L}_{Grav} + \mathcal{L}_\Psi + \mathcal{L}_I$ with that for the Maxwell-Dirac system leading to quantum electrodynamics with the corresponding action defined on the background of Minkowski spacetime. For the sake of convenience, in what follows we keep the notation of previous sections, and continue to use the arbitrary observer field v^α as a difference field between the four-velocity vector field u^α and the arbitrary co-vector field A_α defined by $v^\alpha := u^\alpha - h^{\alpha\beta}A_\beta$. Let us emphasize once again that $A = A_\mu dx^\mu$ appearing in the action $\mathcal{I}$ is simply an arbitrary 1-form on $\mathcal{M}$, as yet bearing no special relation to the 2-form F of equation (2.6).

Extremizations of the action $\mathcal{I}$ with respect to variations of the multiplier fields ζ, Υ_μ, $\Upsilon^\mu_{\nu\sigma}$, $\Upsilon^{\mu\nu}$, and $\tilde{\Upsilon}^{\mu\nu}$ immediately yield the normalization condition

$$u^\nu t_\nu = 1 \tag{2.95}$$

for the timelike vector-field u and the conditions

$$h^{\mu\nu}t_\nu = 0, \quad \nabla_\mu h^{\nu\sigma} = 0, \quad \nabla_\mu t_\nu = 0, \quad \text{and} \quad \partial_{[\mu}t_{\nu]} = 0 \tag{2.96}$$

specifying the Galilean structure (cf. Eq. (2.21)). Whereas its extremization with respect to variations of the tensor-field $\chi^{\mu\nu}$ yields the equation (2.29):

$$2\nabla_{[\mu}A_{\nu]} = -2\overset{u}{h}_{\sigma[\mu}\nabla_{\nu]}u^\sigma. \tag{2.97}$$

Consequently, once the last equation is compared with equation (2.13), we immediately obtain the condition (2.28) entailing that the 2-form F appearing in the connection (2.6) is closed. As discussed in subsection 2.2.2, the condition (2.28) is equivalent to the Newton-Cartan field equation

$$R^{\alpha}{}_{\beta}{}^{\gamma}{}_{\delta} = R^{\gamma}{}_{\delta}{}^{\alpha}{}_{\beta} \,, \tag{2.98}$$

which picks out the Newton-Cartan connection from the general Galilean connections (2.6). Thus, the extremization of $\mathcal{I}$ with respect to $\chi^{\mu\nu}$ not only yields this Newton-Cartan field equation, but, thereby, together with the relations (2.96) and (2.95), also fixes the hitherto unspecified connection to be Newton-Cartan — i.e., the one given by equations (2.24) and (2.30). What is more, as a result of extremizations of the action with respect to variations of the tensor-field $\chi^{\alpha\mu\nu}$ we also have the relation

$$\nabla_{[\gamma}\nabla_{\delta]} A_{\mu} - \overset{u}{h}_{\mu\nu} \nabla_{[\gamma}\nabla_{\delta]} u^{\nu} = 0 \,, \tag{2.99}$$

or, equivalently (cf. Eq. (2.42)),

$$h^{\lambda\sigma} R^{\alpha}{}_{\sigma\gamma\delta} \equiv R^{\alpha\lambda}{}_{\gamma\delta} = 0 \,, \tag{2.100}$$

which, together with (2.98), allows us to identify our spacetime structure $(\mathcal{M}; h, \tau, \nabla)$ as the Newton-Cartan structure (cf. the last paragraph of subsection 2.2.3). Consequently, we can now recognize the arbitrary covector-field A_{μ} as the gravitational 'vector-potential' defined by equation (2.28) and adopt the entire body of formulae exclusive to the Newton-Cartan structure from sections 2.2 and 2.3. In particular, we can now use the expression (2.26) for the curvature tensor associated with the Newton-Cartan connection to obtain the corresponding Ricci tensor

$$R_{\mu\nu} = \{h^{\alpha\beta} \overset{v}{\nabla}_{\alpha}\overset{v}{\nabla}_{\beta} \overset{v}{\Phi}\}t_{\mu\nu} \,, \tag{2.101}$$

which, of course, is gauge-independent despite its appearance. One way to see this gauge-independence is to note that — thanks to the tracelessness of the gravitational field tensor ($\overset{v}{G}_\mu{}^\alpha{}_\alpha = 0$; cf. Eq. (2.25)) — the distinction between the 'flat' covariant derivative $\overset{v}{\nabla}_\alpha$ and the 'curved' covariant derivative ∇_α disappears in the case of divergence. In particular, $\overset{v}{\nabla}_\alpha \overset{v}{\Phi}{}^\alpha = \nabla_\alpha \overset{v}{\Phi}{}^\alpha$, where $\overset{v}{\Phi}{}^\alpha := h^{\alpha\beta} \overset{v}{\nabla}_\beta \overset{v}{\Phi} = h^{\alpha\beta} \partial_\beta \overset{v}{\Phi} = h^{\alpha\beta} \nabla_\beta \overset{v}{\Phi}$ because $\overset{v}{\Phi}$ is a scalar. Therefore, as a direct consequence, the above expression for the Ricci tensor can also be written in terms of the *curved* covariant derivative operator ∇_α as

$$R_{\mu\nu} = \{\nabla_\alpha \overset{v}{\Phi}{}^\alpha\} t_{\mu\nu} = \{h^{\alpha\beta} \nabla_\alpha \nabla_\beta \overset{v}{\Phi}\} t_{\mu\nu} \,. \tag{2.102}$$

To extract further dynamical information from the action $\mathcal{I}$, we next extremize it with respect to variations of the covector-field A_α and look for the corresponding Euler-Lagrange equations

$$\frac{\delta\mathcal{L}}{\delta A_\alpha} \equiv \frac{\partial\mathcal{L}}{\partial A_\alpha} - \nabla_\mu\{\frac{\partial\mathcal{L}}{\partial(\nabla_\mu A_\alpha)}\} + \nabla_\mu\nabla_\nu\{\frac{\partial\mathcal{L}}{\partial(\nabla_\mu\nabla_\nu A_\alpha)}\} = 0\,. \tag{2.103}$$

Since the Newton-Cartan connection Γ is invariant under variations of the gauge variables A and u (cf. subsection 2.2.4), these equations give the relations $\frac{\delta\mathcal{L}_\mathrm{B}}{\delta A_\alpha} = \frac{\delta\mathcal{L}_\mathrm{A}}{\delta A_\alpha} = \frac{\delta\mathcal{L}_\Psi}{\delta A_\alpha} = 0$, and

$$\frac{\delta\mathcal{L}}{\delta A_\alpha} = \wp\,\nabla_\sigma \chi^{\alpha\sigma} + \wp\,\nabla_\sigma\nabla_\gamma \chi^{\alpha\sigma\gamma} + \wp\,\kappa\,v^\alpha\,\nabla_\sigma\Theta^\sigma - \wp\,4\pi G\,J^\alpha = 0\,. \tag{2.104}$$

As it stands, the last equation explicitly contains the observer field v^α, which can be eliminated by contracting both sides of the equation with t_α, and using $t_\alpha v^\alpha = 1$. Subsequently, after using (2.94) with $t_\alpha u^\alpha = 1$ in the resulting equation and dividing it through by $\wp$, we obtain

$$\kappa\,\nabla_\sigma\Theta^\sigma + t_\alpha\nabla_\sigma \chi^{\alpha\sigma} + t_\alpha\nabla_\sigma\nabla_\gamma \chi^{\alpha\sigma\gamma} = 4\pi G\,\rho\,, \tag{2.105}$$

where $\rho \equiv m\Psi\overline{\Psi}$ as in equation (2.75).

Amiably enough, it turns out that extremization of the action with respect to the æther-field u^α leads back to the same equation (2.105); i.e., the dynamical pieces of information extractable from the variations of $\mathcal{I}$ with respect to A_α and u^α are identical. The components of Euler-Lagrange equations in this twin case are: $\frac{\delta \mathcal{L}_B}{\delta u^\alpha} = \frac{\delta \mathcal{L}_A}{\delta u^\alpha} = 0$,

$$\frac{\delta \mathcal{L}_\Gamma}{\delta u^\alpha} = +\wp \left\{ \zeta \, t_\alpha - \overset{u}{h}_{\alpha\mu} \nabla_\gamma \chi^{\mu\gamma} \right\}, \tag{2.106a}$$

$$\frac{\delta \mathcal{L}_N}{\delta u^\alpha} = -\wp \left\{ \overset{u}{h}_{\alpha\mu} \nabla_\delta \nabla_\gamma \chi^{\mu\delta\gamma} \right\}, \tag{2.106b}$$

$$\frac{\delta \mathcal{L}_\Phi}{\delta u^\alpha} = +\wp \, \kappa \, A_\alpha \left\{ h^{\delta\gamma} \nabla_\delta \nabla_\gamma \Theta(A_\mu, u^\nu) \right\}, \tag{7.106c}$$

$$\frac{\delta \mathcal{L}_\Psi}{\delta u^\alpha} = +\wp \, 4\pi G \, i \frac{\hbar}{2} \left\{ \Psi \partial_\alpha \overline{\Psi} - \overline{\Psi} \partial_\alpha \Psi \right\}, \tag{2.106d}$$

and $$\frac{\delta \mathcal{L}_I}{\delta u^\alpha} = -\wp \, 4\pi G \left\{ m\Psi\overline{\Psi} \right\} A_\alpha. \tag{2.106e}$$

$$\tag{2.106}$$

In evaluating the first two of the displayed equations we have used properties (2.12), (2.16), and (2.17) of a geodesic observer. Combining all of these components in the Euler-Lagrange equation $\frac{\delta \mathcal{L}}{\delta u^\alpha} = 0$, dividing it through by $\wp$, contracting it with $h^{\sigma\alpha}$ and substituting $\delta_\mu^\sigma - t_\mu u^\sigma$ in it for $h^{\sigma\alpha} \overset{u}{h}_{\alpha\mu}$, using equation (2.104) to substitute for $\nabla_\gamma \chi^{\sigma\gamma} + \nabla_\delta \nabla_\gamma \chi^{\sigma\delta\gamma}$, and, finally, contracting the result of all these operations (in this order) with t_σ and using $t_\sigma u^\sigma = 1$ yields equation (2.105) as asserted. This result is hardly surprising since, as we shall see in a moment, the momenta canonically conjugate to the variables A_μ and u^ν are directly proportional to each other.

Before we can analyze equation (2.105) any further, however, we need to either eliminate the undetermined multiplier tensors $\chi^{\mu\nu}$ and $\chi^{\alpha\mu\nu}$, or interpret them in physical terms. A physical meaning of the multiplier field $\chi^{\mu\nu}$ is readily revealed if we evaluate the four-momentum density canonically conjugate to the gravitational field variable A_μ at each point x on $\mathcal{M}$:

$$\Pi^\mu \, \delta A_\mu \; := \; \mathcal{J}_A^\alpha \, t_\alpha \; = \; \wp \, (t_\alpha \chi^{\alpha\mu}) \, \delta A_\mu \,, \tag{2.107}$$

where

$$\mathcal{J}_A^\mu \; := \; \{\frac{\delta \mathcal{L}}{\delta (\nabla_\mu A_\alpha)}\} \, \delta A_\alpha \; + \; \{\frac{\delta \mathcal{L}}{\delta (\nabla_\mu \nabla_\nu A_\alpha)}\} \nabla_\nu \, \delta A_\alpha \tag{2.108}$$

is the *presymplectic potential current density* [44, 45], defined and discussed in the appendix below (cf. Eq. (2.219)). Here we have dropped the 4-divergence term

$$\wp \, \nabla_\nu \, (t_\mu \, \chi^{\alpha\mu\nu} \, \delta A_\alpha) \tag{2.109}$$

from the expression (2.107), arising from the $\mathcal{L}_N$ component of the action, since, thanks to the vanishing of the boundary of a boundary theorem, it does not contribute to the *presymplectic potential* (2.222). Note that the antisymmetric nature of $\chi^{\mu\nu}$ requires Π^μ to be spacelike: $t_\mu \Pi^\mu \equiv 0$. It is also noteworthy that the only non-zero contribution to the conjugate momentum density comes from the $\mathcal{L}_\Gamma$ component of the action, which is also responsible for the condition (2.27) specifying the Newton-Cartan connection out of the generic possibilities (2.6). Thus, we now understand the physical meaning of the multiplier field $\chi^{\mu\nu}$. The physical meaning of the multiplier field $\chi^{\alpha\mu\nu}$, on the other hand, has already been anticipated in the definition (2.93) of Λ_N, which, in what follows, will be viewed as a cosmological contribution.

The relations (2.107) and (2.93) allow us to rewrite equation (2.105) in terms of the physical field variables Ψ, $\overline{\Psi}$, u^ν, A_μ, and the canonical conjugate field Π^μ of A_μ:

$$\kappa \, \nabla_\sigma \Theta^\sigma(A_\mu, u^\nu) \; + \; \frac{1}{\wp} \nabla_\sigma \Pi^\sigma \; + \; \Lambda_N \; = \; 4\pi G \, m \Psi \overline{\Psi} \; \equiv \; 4\pi G \, \rho \,, \tag{2.110}$$

where we have made use of $\wp \, t_\alpha \nabla_\sigma \chi^{\alpha\sigma} = \wp \, \nabla_\sigma(t_\alpha \chi^{\alpha\sigma}) = \nabla_\sigma \Pi^\sigma$. It is not surprising that only the four-momentum density Π^μ conjugate to the field variable

A_μ appears in this expression, because, as one may expect, the four-momentum density $\overset{u}{\Pi}_\nu$ conjugate to the observer field u^ν is just the 'anti-dual' of Π^μ,

$$\overset{u}{\Pi}_\nu \, \delta u^\nu \; := \; \mathcal{J}_u^\alpha \, t_\alpha \; = \; (- \, \overset{u}{h}_{\nu\mu} \, \Pi^\mu) \, \delta u^\nu \, , \tag{2.111}$$

(again, modulo the 4-divergence term

$$\wp \, \nabla_\nu \, (t_\mu \, \chi^{\alpha\mu\nu} \, \overset{u}{h}_{\alpha\sigma} \, \delta u^\sigma) \tag{2.112}$$

not affecting the symplectic structure) which can be easily checked by explicitly evaluating $\mathcal{J}_u^\alpha$. In fact, we have the relationship

$$\Pi^\mu \; + \; h^{\mu\nu} \, \overset{u}{\Pi}_\nu \; = \; 0 \; = \; \overset{u}{\Pi}_\mu \; + \; \overset{u}{h}_{\mu\nu} \, \Pi^\nu \, , \tag{2.113}$$

which indicates that, strictly speaking, we should not view A_μ and u^ν as independent variables. In what follows, however, for the sake of convenience, we shall continue treating them as if they were independent. Eventually, in the next section, the constraint (2.113) will be taken into account in a consistent manner.

Now, from the extremization of the action with respect to variations of the scalar multiplier field χ we immediately infer that

$$\frac{1}{\wp} \nabla_\sigma \Pi^\sigma \; - \; (1 - \kappa) \, \nabla_\sigma \Theta^\sigma \; + \; \lambda \, \Lambda_{\text{N}} \; - \; \Lambda_{\text{O}} \; = \; 0 \, , \tag{2.114}$$

which, upon substitution into equation (2.110), yields

$$\nabla_\sigma \Theta^\sigma \; + \; \Lambda \; = \; 4\pi G \, \rho \, , \tag{2.115}$$

where

$$\Lambda \; := \; \Lambda_{\text{O}} + (1 - \lambda)\Lambda_{\text{N}} \, . \tag{2.116}$$

For the flat Galilean spacetime, the left-hand side of this equation is simply the Laplacian of the scalar $\Theta(A_\mu, \, u^\nu)$ plus the cosmological parameter Λ; and, hence, it is just the Poisson equation of the classical Newtonian theory of gravity provided we can interpret $\Theta(A_\mu, u^\nu)$ as the corresponding scalar gravitational potential. But, of course, with A_μ recognized to be the gravitational vector-potential as a result of equations (2.95), (2.96), and (2.98), it is indeed possible to identify the function $\Theta(A_\mu, u^\nu)$ in equation (2.115) with the Newtonian gravitational scalar-potential $\overset{v}{\Phi}(A_\alpha, u^\nu)$ with respect to the observer-field $v^\alpha \equiv u^\alpha - h^{\alpha\beta}A_\beta$:

$$\overset{v}{\Phi}(A_\alpha, u^\nu) \equiv \Theta(A_\alpha, u^\nu) = \frac{1}{2} h^{\mu\nu} A_\mu A_\nu - A_\sigma u^\sigma \qquad (2.117)$$

(cf. Eq. (2.32)). Using this identification in equation 2.102, and multiplying equation (2.115) with $t_{\mu\nu}$, it is now easy to obtain the last of the Newton-Cartan field equations,

$$R_{\mu\nu} + \Lambda\, t_{\mu\nu} = 4\pi G\, M_{\mu\nu}, \qquad (2.118)$$

as a generally-covariant generalization of the Newton-Poisson equation. As noted before, an immediate inference one gains from this field equation is that spacelike hypersurfaces of simultaneity Σ_t embedded in $\mathcal{M}$ are copies of flat, Euclidean three-spaces: $h^{\mu\alpha}h^{\nu\sigma}R_{\alpha\sigma} = 0$. Unlike the general relativistic case, however, here the nonrelativistic contracted Bianchi identities (2.20) by themselves do not render the scalar Λ to be a spacetime constant. Nevertheless, given the condition (2.98) on $\mathcal{M}$, a detailed analysis [10] of the tensor representations of the Galilean group provided by both the physically sensible matter tensor $M^{\mu\nu}$ and the curvature tensor $R^\alpha{}_{\mu\sigma\nu}$ of the general Galilean spacetime, together with the constraints due to nonrelativistic contracted Bianchi identities (2.20), reveals that Λ in the above field equation at the most could be a spacetime constant.

Note that neither Λ_{O} nor Λ_{N} are individually required to be spacetime constants, only the net Λ is. Further, there is nothing sacrosanct about the interpretation we have given to Λ_{N}. This contribution to Λ arises from the $\mathcal{L}_{\mathrm{N}}$ term in the action. This is the term which gives rise to the additional constraint (2.41) on the curvature tensor. But, as discussed in subsection 2.2.3, in the asymptotic limit this constraint is automatically satisfied, and, hence, the $\mathcal{L}_{\mathrm{N}}$ term can be dropped from the action; i.e, in such a case, the variation of the multiplier tensor $\chi^{\alpha\mu\nu}$ itself must be set equal to zero. Moreover, in this limit any cosmological contribution is also generally ruled out. And, indeed, these physical requirements all come out consistently in our interpretation of Λ_{N} if we set $\lambda = \frac{\Lambda_{\mathrm{O}}}{\Lambda_{\mathrm{N}}}$ in the Lagrangian $\mathcal{L}_{\Lambda}$ giving $\Lambda \equiv \Lambda_{\mathrm{N}}$. However, if one wishes, one can easily set $\lambda = 1$ to avoid such a strong link between the 'Newtonian restriction' (2.41) and the cosmological constant. As long as the controversy over cosmological contributions is unsettled, a choice of this part of the action is simply a matter of taste.

Turning now to the matter part of the action, we recognize the Lagrangian density $\mathcal{L}_{\Psi} + \mathcal{L}_{\mathrm{I}}$ as nothing but the Schrödinger-Kuchař Lagrangian density $\mathcal{L}_{Kuch}$ expressed by equation (2.72) of the previous section. Since $\mathcal{L}_{Grav}$ is independent of the variables Ψ, $\partial_{\mu}\Psi$, and their conjugates, extremizations of the action $\mathcal{I}$ with respect to variations of Ψ and $\overline{\Psi}$ are identical to those of an action with Lagrangian density $\mathcal{L}_{Kuch} \equiv \mathcal{L}_{\Psi} + \mathcal{L}_{\mathrm{I}}$ which we have already discussed in that section. The resulting Euler-Lagrange equations $\frac{\delta\mathcal{L}}{\delta\overline{\Psi}} = 0$ and $\frac{\delta\mathcal{L}}{\delta\Psi} = 0$ yield, respectively, the Schrödinger-Kuchař equation (2.74),

$$[\frac{\hbar^2}{2m}\nabla^{\alpha}\partial_{\alpha} + i\hbar(u^{\alpha} - h^{\alpha\beta}A_{\beta})\partial_{\alpha} + m(u^{\alpha}A_{\alpha} - \tfrac{1}{2}h^{\alpha\beta}A_{\alpha}A_{\beta}) + i\frac{\hbar}{2}\nabla_{\alpha}(u^{\alpha} - h^{\alpha\beta}A_{\beta})]\Psi = 0,$$

(2.119)

and its conjugate, describing the motion of a classical Galilean-relativistic Schrödinger-Kuchař field Ψ on the curved Newton-Cartan manifold achieved by minimally coupling it to the gravitational 'vector-potential' A_{μ}.

Finally, let us not forget the remaining kinematical variables $^{(s)}y : \mathcal{M} \to \overset{\circ}{\mathcal{M}}$. Extremization of the action under variations of these auxiliary variables leads to

$$0 = \frac{d}{ds}\mathcal{I} \equiv \delta\mathcal{I} = \int \frac{\delta\mathcal{I}_{Grav}}{\delta^{(s)}y^{\alpha}}\,\delta^{(s)}y^{\alpha} + \int \frac{\delta\mathcal{I}_{Kuch}}{\delta^{(s)}y^{\alpha}}\,\delta^{(s)}y^{\alpha}.$$

(2.120)

As we already saw in the subsection 2.2.5 and the section 2.3, given the matter field equations (2.119) and its complex conjugate, one of the consequences of the covariance of the matter action $\mathcal{I}_{Kuch}$ is the matter conservation laws

$$\nabla_\mu J^\mu = 0 \quad \text{and} \quad \nabla_\mu M^{\mu\nu} = 0 \tag{2.121}$$

(cf. Eqs. (2.64a) and (2.66)), which correspond to the relativistic conservation law $\nabla_\mu T^{\mu\nu} = 0$. In addition, the covariance of the matter action reduces equation 2.120 to

$$0 = \int \frac{\delta \mathcal{I}_{Grav}}{\delta^{(s)} y^\alpha} \delta^{(s)} y^\alpha = \int \frac{\delta \mathcal{I}_{Grav}}{\delta h^{\mu\nu}} \delta h^{\mu\nu} + \int \frac{\delta \mathcal{I}_{Grav}}{\delta t_\mu} \delta t_\mu + \int \frac{\delta \mathcal{I}_{Grav}}{\delta A_\mu} \delta A_\mu$$
$$+ \int \frac{\delta \mathcal{I}_{Grav}}{\delta u^\nu} \delta u^\nu , \tag{2.122}$$

since the variations of the action $\mathcal{I}_{Grav}$ with respect to all of the multiplier fields have been required to vanish. But now we can parallel the reasoning of the subsection 2.2.5 with $\mathcal{I}_m$ replaced by $\mathcal{I}_{Grav}$ and obtain the condition

$$\nabla_\mu \mathcal{G}^{\mu\nu} = 0 \tag{2.123}$$

analogous to the equation (2.121) above, where

$$\mathcal{G}^{\mu\nu} := \{\nabla_\sigma \Theta^\sigma + \Lambda\} u^\mu u^\nu - 2\{\nabla_\sigma \Theta^\sigma + \Lambda\} u^{(\mu} h^{\nu)\sigma} A_\sigma - h^{\mu\sigma} h^{\nu\alpha}(S_{Grav})_{\sigma\alpha} \tag{2.124}$$

(cf. Eqs. (2.67) and (2.115)). Comparing this latter expression with the general definition (2.37) for $R^{\mu\nu}$ we immediately see that it is equivalent to

$$\mathcal{G}^{\mu\nu} \equiv R^{\mu\nu} + \Lambda v^\mu v^\nu . \tag{2.125}$$

Substituting this expression into equation (2.123) and using $\nabla_\sigma v^\sigma = 0 = v^\mu \nabla_\mu v^\alpha$ for the geodetic observer v^α gives

$$\nabla_\mu R^{\mu\nu} = 0 , \tag{2.126}$$

63

which is nothing but the contracted Bianchi identities (2.20) because $h^{\mu\nu}R_{\mu\nu} =: R = 0$ according to equation (2.118). Thus, in close analogy with the Einstein-Hilbert theory, extremization of the total action with respect to variations of the supplementary variables $^{(s)}y$ yields both the matter conservation law and the contracted Bianchi identities.

It is remarkable that we have been able to obtain *all* of the field equations of the Newton-Cartan theory, (2.43), the matter conservation laws (2.64a) and (2.66), as well as the equation of motion (2.74) for the classical Schrödinger-Kuchař field Ψ on the curved Newton-Cartan spacetime, from a *single* action principle[6].

(6. All previous attempts to satisfactorily reformulate Newton-Cartan theory in four-dimensions using variational principles have met insurmountable difficulties due to the non-semi-Riemannian nature of the Galilean spacetime. For a partially successful attempt, see reference [47]. For a review of a five-dimensional formulation overcoming at least some of the difficulties, see reference [48]).

What is more, the resultant theory we have obtained is covariant under the complete gauge group $\mathcal{A}ut(B(\mathcal{M}))$ discussed in the subsection 2.2.4. The invariance of $\mathcal{L}_{Kuch}$ under this gauge group has already been emphasized in the previous section. Among the four components of the Lagrangian density associated with the connection-field, it can be explicitly checked that $\mathcal{L}_{\mathrm{B}}$, $\mathcal{L}_{\Gamma}$, $\mathcal{L}_{\mathrm{N}}$, and $\mathcal{L}_{\Lambda}$ are all invariant under the complete automorphism group $\mathcal{A}ut(B(\mathcal{M}))$ (the factor $\{h^{\sigma\gamma}\nabla_{\sigma}\nabla_{\gamma}\Theta\}$, of course, does not change under the full gauge transformation; (cf. Eqs. (2.101) and (2.102)). The third one, $\mathcal{L}_{\Phi}$, on the other hand, is invariant only under the diffeomorphism subgroup, $\mathrm{Diff}(\mathcal{M})$, and the 'boost' subgroup,

$$u^{\alpha} \mapsto u^{\alpha} + h^{\alpha\sigma}\mathrm{w}_{\sigma}\,,$$

$$A_\alpha \mapsto A_\alpha + \mathrm{w}_\alpha - (u^\sigma \mathrm{w}_\sigma + \tfrac{1}{2} h^{\mu\nu} \mathrm{w}_\mu \mathrm{w}_\nu) t_\alpha \,, \tag{2.127}$$

of the full group $\mathcal{A}ut(B(\mathcal{M}))$. It is not invariant, in particular, under the transformations $A_\mu \mapsto A_\mu + \partial_\mu f$. As a result, the Euler-Lagrange equation (2.104) depends on the choice of this internal gauge; for, under these transformations, $v^\mu \mapsto v^\mu - h^{\mu\nu} \partial_\nu f$. However, this gauge-dependence is projected out in the actual field equation (2.105) — since the gauge-dependent part is spacelike: $t_\alpha \partial^\alpha f = 0$ — making the whole theory invariant under the complete Newton-Cartan gauge group $\mathcal{A}ut(B(\mathcal{M}))$. Better still, if one wishes — say, for aesthetic reasons — to eliminate the gauge-dependent part $\mathcal{L}_\Phi$ of the total Lagrangian density to make the Lagrangian prescription *manifestly* generally-covariant, one is completely free to do so by simply choosing the arbitrary constant $\kappa = 0$. Therefore, our variational reformulation of the classical Newton-Cartan-Schrödinger theory is no less generally-covariant than Einstein's theory of gravity (modulo, of course, the philosophical caveat made in the note 2). In fact, in close analogy with variational formulations of Einstein's theory, all we have assumed here *a priori* is an $\mathcal{A}ut(B(\mathcal{M}))$-invariant action functional — dependent only on local degrees of freedom — defined on some measurable region of a sufficiently smooth, real, differentiable Hausdorff 4-manifold $\mathcal{M}$. The rest follows squarely from extremizations of this action functional.

2.4.2 Constraint-free Hamiltonian formulation in 3+1 dimensions

Unlike the spacetime covariant Lagrangian formulation of any field theory, the conventional Hamiltonian formulation of such a theory requires a blatantly non-covariant Aristotelian 3+1 decomposition of spacetime into 3-spaces Σ_t at instants of time $t \in \mathbb{R}$. For such a foliation of spacetime to be possible, it is necessary to assume that the manifold $\mathcal{M}$ is of the globally hyperbolic form:

$\mathcal{M} = \mathbb{R} \times \Sigma$, where Σ is taken to be an embedded (cf. note 3) achronal[7] closed submanifold of $\mathcal{M}$ such that its domain of dependence $\mathcal{D}(\Sigma) = \mathcal{M}$.

(7. A set is achronal if none of its elements can be joined by a timelike curve. For a definition of 'the domain of dependence' and a general discussion on the construction of regular Cauchy surfaces, see reference [28].)

In the general relativistic case such an *a priori* topological constraint on spacetime is obviously too severe, and therefore canonical approaches to quantize Einstein's gravity are sometimes criticized for being overly restrictive if not completely misguided. However, in Newtonian physics time plays a very privileged role. Therefore, for a Galilean spacetime the breakup $\mathcal{M} = \mathbb{R} \times \Sigma$ is not only natural but, in fact, a part of the intrinsic structure determined by the non-dynamical metrics $h^{\mu\nu}$ and $t_{\mu\nu}$. This fact, of course, does not compensate for the loss of covariance of any field theory based on such a breakup of spacetime into 3-spaces at times. However, as we shall see in the next subsection, the apparent loss of covariance in the Hamiltonian formulation of our theory is only an aesthetic loss, and can be rectified using the relatively less-popular symplectic approach yielding a manifestly covariant description of canonical formalism [44, 45, 49–51].

To see in detail the intuitively obvious fact that Newton-Cartan structure is naturally globally hyperbolic, first recall that the structure $(\mathcal{M}, t_{\mu\nu})$ is time-orientable; i.e., there exists a globally defined smooth vector field t_μ on $\mathcal{M}$ inducing the temporal metric $t_{\mu\nu} = t_\mu t_\nu$ and determining a time-orientation. Further, since $\mathcal{M}$ is contractible by definition, the compatibility condition $\nabla_\mu t_\nu = 0$ together with the Poincaré lemma allows one to define the absolute time globally by a map $t : \mathcal{M} \to \mathbb{R}$, foliating the spacetime *uniquely* into one-parameter family of smooth Cauchy surfaces Σ_t — the domains of simultaneity. Here, one can change the scalar function t into $t' = t'(t)$, but the regular foliation $\mathcal{M} = \cup \{\Sigma_t\}$ with $\Sigma_t = \{x \in \mathcal{M} \,|\, t(x) = c \,, c \in \mathbb{R}\}$ remains a part of the intrinsic structure of spacetime. Moreover, these Cauchy surfaces have the property that all curves with images confined to them are spacelike; consequently, they may also be defined directly by the 1-form $\tau = t_\alpha dx^\alpha$ as the 3-dimensional subspaces [11]

$$\Sigma_x := \{\eta \in \mathrm{T}_x\mathcal{M} \mid \eta \lrcorner \tau = 0\} \tag{2.128}$$

of tangent spaces $\mathrm{T}_x\mathcal{M}$ at points x on $\mathcal{M}$. Since τ is closed, this differential system of hypersurfaces Σ_x is completely integrable, and defines a foliation of regular Cauchy surfaces on $\mathcal{M}$ which are given by Σ_t with a locally defined scalar function $t(x) \in C^\infty(\mathcal{M})$ satisfying $dt = \tau$. What is more, these smooth 3-surfaces Σ_t are orientable since the 4-manifold $\mathcal{M}$ is orientable; given the 4-volume measure $\varepsilon_{\mu\nu\rho\sigma}$ on $\mathcal{M}$ (cf. Eq. 2.3), the corresponding contravariant tensor $\varepsilon^{\mu\nu\rho\sigma}$ can be used [29] to obtain a 3-volume measure $\varepsilon^{[\mu\nu\rho]} = \varepsilon^{\mu\nu\rho} := \varepsilon^{\mu\nu\rho\sigma} t_\sigma$ on the Cauchy surfaces Σ_t. In the special case of Newton-Cartan spacetime these Cauchy surfaces are flat Riemannian 3-surfaces of spacelike vectors, carrying a non-degenerate Euclidean metric induced by projecting $h^{\mu\nu}$ with respect to any unit timelike vector field — say the Galilean observer field u^α. That is, $(\Sigma_t, \overset{u}{h}_{\mu\nu})$, with $\overset{u}{h}_{\mu\nu}$ as the induced metric field on Σ_t corresponding to the vector field u^α, is a copy of Euclidean 3-space, and, hence, $\mathcal{M}$ is homeomorphic to $\mathbb{R}^4$. Since $\overset{u}{h}_{\mu\nu}$ has a spatial inverse, namely $h^{\mu\nu}$ (cf. Eq. (2.9)), it follows that [13] there exists a unique 3-dimensional derivative operator $^{(3)}\nabla_\mu$ on Σ_t compatible with $\overset{u}{h}_{\mu\nu}$: $^{(3)}\nabla_\sigma \overset{u}{h}_{\mu\nu} = 0$. It can be characterized in terms of the 4-dimensional operator ∇_μ; for example, for a tensor field $V^\alpha_{\mu\nu}$,

$$^{(3)}\nabla_\sigma V^\alpha_{\mu\nu} = \overset{u}{\delta_\sigma}{}^\beta \overset{u}{\delta_\mu}{}^\gamma \overset{u}{\delta_\nu}{}^\lambda \nabla_\beta V^\alpha_{\gamma\lambda} . \tag{2.129}$$

Note that there is no need to project the contravariant indices because they remain spacelike even after the application of ∇_μ because of the compatibility condition $\nabla_\mu t_\nu = 0$, and that $^{(3)}\nabla_\sigma$ satisfies

$$^{(3)}\nabla_\sigma \overset{u}{\delta}_\mu{}^\nu \;=\; ^{(3)}\nabla_\sigma \overset{u}{h}_{\mu\nu} \;=\; ^{(3)}\nabla_\sigma h^{\mu\nu} \;=\; 0 \qquad (2.130)$$

over and above all of the defining conditions for a derivative operator. The two operators ∇_σ and $^{(3)}\nabla_\sigma$ induce the same parallel transport condition for spacelike vectors on Σ_t : $U^\alpha (^{(3)}\nabla_\alpha)V^\mu = U^\alpha \overset{u}{\delta}_\alpha{}^\gamma \nabla_\gamma V^\mu = U^\gamma \nabla_\gamma V^\mu$, for all spacelike fields U^μ and V^μ on Σ_t.

The vector field u^α may be viewed as describing the flow of time in $\mathcal{M}$ and can be used to identify each Cauchy surface Σ_t with the initial surface Σ_0. Given such an observer field u^α and its relative spatial projection field $\overset{u}{\delta}_\mu{}^\nu \equiv \overset{u}{h}_{\mu\sigma} h^{\sigma\nu}$, one may decompose any spacetime quantity into its projections normal and tangential to Σ_t. For example, using the relations

$$t_\alpha \overset{u}{\delta}_\mu{}^\alpha \;=\; 0 \quad \text{and} \quad u^\alpha \overset{u}{\delta}_\alpha{}^\mu \;=\; 0 , \qquad (2.131)$$

a contravector V^α on $\mathcal{M}$ may be decomposed as

$$V^\alpha \;=\; {}^\perp V \, u^\alpha \;+\; {}^\parallel V^\alpha \qquad (2.132)$$

with $^\perp V := t_\mu V^\mu$ and $^\parallel V^\alpha := \overset{u}{\delta}_\mu{}^\alpha V^\mu$, whereas a covector V_α on $\mathcal{M}$ can be split as

$$V_\alpha \;=\; {}_\perp V \, t_\alpha \;+\; {}_\parallel V_\alpha \qquad (2.133)$$

with $_\perp V := u^\mu V_\mu$ and $_\parallel V_\alpha := \overset{u}{\delta}_\alpha{}^\mu V_\mu$. Note that the distinction between covectors and contravectors made conspicuous by these decompositions has much greater significance here than in the general relativistic spacetimes.

With the above rather elaborate delineation of the convenient fact that Newton-Cartan spacetime is naturally globally hyperbolic, we are ready to cast our theory in a Hamiltonian form. The next two steps for which are, firstly, to construct a configuration space for the field variables on $\mathcal{M}$ by specifying instantaneous configuration fields corresponding to these variables on a chosen Cauchy surface Σ_t, and then, secondly, to evaluate conjugate momentum densities of these instantaneous fields on the chosen surface. As we shall see, the ensuing canonical variables constitute the proper Cauchy data to be propagated from one Cauchy surface to another. Let us begin by concentrating on the pair (A_μ, u^ν) of pure gravitational field variables on $\mathcal{M}$ and tentatively take A_μ and u^ν evaluated on Σ_t as our instantaneous configuration variables. Using the defining equations (2.133) and (2.132), these variables may be decomposed into their components normal and tangential to Σ_t as

$$A_\alpha \;=\; {}_\perp A\, t_\alpha \;+\; {}_\parallel A_\alpha \tag{2.134}$$

with ${}_\perp A := u^\mu A_\mu$ and ${}_\parallel A_\alpha := \overset{u}{\delta}{}_\alpha{}^\mu A_\mu$, and

$$u^\alpha \;=\; {}^\perp u\, u^\alpha \;+\; {}^\parallel u^\alpha \tag{2.135}$$

with ${}^\perp u := t_\mu u^\mu = 1$ and ${}^\parallel u^\alpha := \overset{u}{\delta}{}_\mu{}^\alpha u^\mu = 0$. The Lagrangian density, expression (2.92), can now be rewritten in terms of these decompositions and normal and tangential parts of the momentum densities conjugate to the fields A_μ and u^ν specified on Σ_t can be evaluated from it. It is easier, however, to use the already evaluated four-momentum densities (2.107) and (2.111) and decompose them according to equations (2.132) and (2.133). Whichever procedure is used, the outcomes are:

$$^\perp\Pi \;=\; t_\mu \Pi^\mu \;=\; 0 \qquad \text{and} \qquad {}^\parallel\Pi^\alpha \;=\; \overset{u}{\delta}{}_\mu{}^\alpha\, \Pi^\mu \;=\; \Pi^\alpha, \tag{2.136}$$

whereas

$$_\perp\overset{u}{\Pi} \;=\; u^\nu\, \overset{u}{\Pi}_\nu \;=\; 0 \qquad \text{and} \qquad {}_\parallel\overset{u}{\Pi}_\alpha \;=\; \overset{u}{\delta}{}_\alpha{}^\mu\, \overset{u}{\Pi}_\mu \;=\; -\,\overset{u}{h}_{\alpha\sigma}\Pi^\sigma. \tag{2.137}$$

The vanishing of the canonical momentum densities $^\perp\Pi$ and $_\perp\overset{u}{\Pi}$ implies that the conjugate momenta are spacelike, and, more importantly for our purposes, suggests that we should not view their conjugate variables $_\perp A$ and $^\perp u$ as dynamical variables because they do not constitute suitable Cauchy data (their presence as dynamical variables would be an obstacle to the Legendre transformation required to specify a Hamiltonian functional on Σ_t). Since $^\perp u \equiv 1$ and $^\parallel u^\alpha \equiv 0$, we shall see that taking the observer field u^ν as one of our configuration variables does not cause any serious problems. However, we must take only the tangential component $_\parallel A_\mu$ of A_μ as our gravitational configuration variable along with the observer field u^ν if we are to maintain substantive uniqueness of the propagation from Cauchy data. In addition to these gravitational field variables, we also have the matter configuration variables Ψ and $\overline{\Psi}$ specified on Σ_t, and, together with them, we also must add their respective conjugate momentum densities

$$\overline{P} := \frac{\delta \mathcal{L}_{Kuch}}{\delta(u^\sigma \partial_\sigma \Psi)} = -2\,i\,\hbar\,\wp\,\pi\,G\,\overline{\Psi} \quad \text{and} \quad P := \frac{\delta \mathcal{L}_{Kuch}}{\delta(u^\sigma \partial_\sigma \overline{\Psi})} = +2\,i\,\hbar\,\wp\,\pi\,G\,\Psi \tag{2.138}$$

to our list of canonical phase-space variables, where $u^\sigma \nabla_\sigma$ is a propagation covariant derivative (i.e., a time derivative) along the unit timelike vector field u^σ.

Next, in order to remain in close contact with the parameterized formulation [38, 46] used in the previous subsection, we view the foliation e of $\mathcal{M}$,

$$e \,:\, \mathbb{R} \times \Sigma \longrightarrow \mathcal{M}\,, \tag{2.139}$$

as the map which allows us to pass to the Hamiltonian formulation of the theory. Then, for each $t \in \mathbb{R}$, e becomes an embedding

$$^{(t)}e \,:\, \Sigma \longrightarrow \mathcal{M} \tag{2.140}$$

(cf. footnote 3). In order to make the embedded hypersurfaces Σ_t (i.e., the images of Σ under the above embedding) spacelike, we require the differential $^{(t)}e_i{}^\mu$ of the map $^{(t)}e$ to satisfy

$$^{(t)}e_i{}^\mu \, t_\mu \;=\; 0 \qquad \text{and} \qquad {}^{(t)}e^i{}_\nu \, u^\nu \;=\; 0 \,, \qquad (2.141)$$

where $^{(t)}e^i{}_\nu$ is the inverse of the differential map $^{(t)}e_i{}^\nu$, and in these expressions (and only in these expressions) the tensors evaluated on the closed 3-surfaces Σ_t are represented via Latin indices for clarity. Comparing these conditions with the relations (2.131), we immediately see that

$$^{(t)}e_i{}^\mu \;\equiv\; \overset{u}{\delta}_i{}^\mu \qquad \text{and} \qquad {}^{(t)}e^i{}_\nu \;\equiv\; \overset{u}{\delta}_\nu{}^i \,. \qquad (2.142)$$

In analogy with what we have done in the previous subsection, we can now pull-back all the fields from $\overset{\circ}{M}$ to the 3-manifold Σ, and enlarge the configuration space of these fields by spacelike embedding variables

$$^{(o)}y \circ {}^{(t)}e \;=:\; \vartheta \,:\, \Sigma \;\longrightarrow\; \overset{\circ}{M} \,, \qquad (2.143)$$

with $^{(o)}y \in \{^{(s)}y\}$ being any fixed diffeomorphism defined by equation (2.89) of the previous subsection. In terms of this map the traditional 'deformation vector' $\dot{\vartheta}^\sigma := u^\sigma \nabla_\sigma \vartheta^\sigma = \frac{\partial \vartheta^\sigma}{\partial t}$ dictating transition from one leaf of foliation to another neighboring one, for example, may be split into its 'lapse' and 'shift' coefficients as

$$\dot{\vartheta}^\sigma \;=\; {}^\perp\!\dot{\vartheta}\, u^\sigma + {}^\|\dot{\vartheta}^\sigma \,, \qquad (2.144)$$

where $^\perp\!\dot{\vartheta} := \dot{\vartheta}^\sigma t_\sigma$ and $^\|\dot{\vartheta}^\sigma := \overset{u}{\delta}_\mu{}^\sigma \dot{\vartheta}^\mu$.

These considerations finally allow reconstruction of the action functional $\mathcal{I}$ of the previous subsection in the Hamiltonian form:

$$\mathcal{I}_{\mathbb{H}} \;=\; \int_{\mathbb{R}} dt \int_{\Sigma_t} d^3x \,\left[{}^\|\Pi^\sigma {}_\|\dot{A}_\sigma + \overset{u}{\Pi}_\sigma \dot{u}^\sigma + \overline{P}\,\dot{\Psi} + P\,\dot{\overline{\Psi}} + \Pi_\sigma \dot{\vartheta}^\sigma - \aleph^\sigma H_\sigma \right], \;\; (2.145)$$

where '·' indicates the time derivative '$u^\sigma \nabla_\sigma$' with respect to the parameter t of the one-parameter family of embeddings, Π_σ are the kinematical momentum densities conjugate to the supplementary embedding variables ϑ^σ, $\dot{\vartheta}^\sigma$ being geometrically a vector field on $\overset{\circ}{\mathcal{M}}$ makes Π_σ a 1-form density of weight one on $\overset{\circ}{\mathcal{M}}$, $\aleph^\sigma$ is a Lagrange multiplier field, $\aleph^\sigma H_\sigma$ defined by

$$\aleph^\sigma[\Pi_\sigma + \overset{\circ}{H}_\sigma] =: \aleph^\sigma H_\sigma \equiv H := {}^{\|}\Pi^\sigma {}_{\|}\dot{A}_\sigma + \overset{u}{\Pi}_\sigma \dot{u}^\sigma + \overline{P}\dot{\Psi} + P\dot{\overline{\Psi}} + \Pi_\sigma \dot{\vartheta}^\sigma - \mathcal{L}$$

(2.146)

(with $\overset{\circ}{H}_\alpha$ being a functional of the original 'un-parameterized' canonical data as well as ϑ^σ representing the combined gravitational and material 'energy-momentum' flux through the surface element $\wp\, u^\alpha$ of the embedding) is the Hamiltonian density giving the Hamiltonian functional,

$$\mathbb{H} = \int_{\Sigma_t} \aleph^\sigma H_\sigma \, d^3x \,,$$

(2.147)

which governs the dynamical evolution of the system, the Lagrangian density $\mathcal{L}$ is given by equation (2.92) with the fields evaluated on Σ_t, and $\wp\, d^3x$ is the volume element in Σ_t. As always, the generalized velocities appearing both explicitly and implicitly in these expressions are viewed as functions of the phase-space variables only. The tensor field $\chi^{\mu\nu} \equiv {}^{\|}\chi^{\mu\nu}$ appearing in $\mathcal{L}$, however, is viewed *not* as a multiplier field, but simply as a function of the canonical variables ${}^{\|}\Pi^\mu$ (cf. Eq. (2.107)). The variables Υ_μ, $\Upsilon^\mu_{\nu\sigma}$, $\Upsilon^{\mu\nu}$, $\tilde{\Upsilon}^{\mu\nu}$, $\aleph^\sigma$, ${}_\perp A$, ζ, $\chi^{\alpha\mu\nu}$, and χ, on the other hand, *are* viewed as non-dynamical multiplier variables, and, hence, the Hamiltonian equations of motion,

$$\frac{\delta H}{\delta \Pi_\alpha} = +u^\sigma \nabla_\sigma \vartheta^\alpha \,, \qquad \frac{\delta H}{\delta \vartheta^\alpha} = -u^\sigma \nabla_\sigma \Pi_\alpha \,,$$

(2.148)

$$\frac{\delta H}{\delta\, {}_{\|}A_\alpha} = -u^\sigma \nabla_\sigma {}^{\|}\Pi^\alpha \,, \qquad \frac{\delta H}{\delta\, u^\alpha} = -u^\sigma \nabla_\sigma \overset{u}{\Pi}_\alpha \,,$$

(2.149)

$$\frac{\delta H}{\delta \, \|\Pi^\alpha} \;=\; +\, u^\sigma \nabla_\sigma \,{}_\|A_\alpha \,, \qquad\qquad \frac{\delta H}{\delta \, \overset{u}{\Pi}_\alpha} \;=\; +\, u^\sigma \nabla_\sigma \, u^\alpha \,, \qquad (2.150)$$

$$\frac{\delta H}{\delta \, \overline{\Psi}} \;=\; -\, u^\sigma \nabla_\sigma \, \mathrm{P} \,, \qquad\qquad \frac{\delta H}{\delta \, \mathrm{P}} \;=\; +\, u^\sigma \nabla_\sigma \, \overline{\Psi} \,, \qquad (2.151)$$

and complex conjugates of the latter two, must be supplemented by the equations

$$\frac{\delta H}{\delta \, \Upsilon_\mu} = 0, \qquad \frac{\delta H}{\delta \, \Upsilon^\mu_{\nu\sigma}} = 0, \qquad \frac{\delta H}{\delta \, \Upsilon^{\mu\nu}} = 0, \qquad \frac{\delta H}{\delta \, \widetilde{\Upsilon}^{\mu\nu}} = 0, \qquad (2.152)$$

$$\frac{\delta H}{\delta \, \aleph^\sigma} = 0, \qquad \frac{\delta H}{\delta \, {}_\perp A} = 0, \qquad \frac{\delta H}{\delta \, \zeta} = 0, \qquad \frac{\delta H}{\delta \, \chi^{\alpha\mu\nu}} = 0, \qquad \text{and} \qquad \frac{\delta H}{\delta \, \chi} = 0, \qquad (2.153)$$

giving rise to constraints on the phase-space. Note that, since $\delta \, {}^\perp u \equiv 0$, no additional constraint corresponding to ${}^\perp u$ need be appended to the set of the Hamiltonian equations of motion. On the other hand, we do have to add the linear constraints

$$\Pi^\mu + h^{\mu\nu} \, \overset{u}{\Pi}_\nu \;=\; 0$$
$$\text{or} \qquad \Pi_\mu + \overset{u}{h}_{\mu\nu} \, \Pi^\nu \;=\; 0 \qquad\qquad (2.154)$$

$$\text{and} \qquad \overline{\mathrm{P}} + 2 i \hbar \, \wp \pi \, G \, \overline{\Psi} \;=\; 0$$
$$\text{or} \qquad \mathrm{P} - 2 i \hbar \, \wp \pi \, G \, \Psi \;=\; 0 \qquad\qquad (2.155)$$

to the above list (cf. Eqs. (2.113) and (2.138)). Here we do not bother to classify these constraints *à la* Dirac [38] because our eventual goal is to get rid of *all* of them in order to obtain a *constraint-free* phase-space.

The first four of these eleven constraint equations, (2.152), immediately yield the conditions (2.96) specifying the Galilean structure. Among the remaining equations, the fifth, $\frac{\delta H}{\delta \aleph^\sigma} = 0$, imparts the kinematical constraint

$$H_\sigma \;\equiv\; \Pi_\sigma + \overset{\circ}{H}_\sigma \;=\; 0 \qquad\qquad (2.156)$$

due to the parameterization process, the sixth, $\frac{\delta H}{\delta_\perp A} = 0$, gives the equation (2.110), the seventh, $\frac{\delta H}{\delta \zeta} = 0$, reproduces the normalization condition $u^\nu t_\nu = 1$ defining the foliation, the eighth, $\frac{\delta H}{\delta \chi^{\alpha\mu\nu}} = 0$, leads to the constraint (2.100) prohibiting rotational holonomy, and the ninth, $\frac{\delta H}{\delta \chi} = 0$, asserts the constraint (2.114).

Among the equations of motion, the first one of the equations (2.148) tells us that the multiplier field $\aleph^\sigma$ is nothing but the generalized velocity field $\dot{\vartheta}^\sigma$,

$$\aleph^\sigma = \dot{\vartheta}^\sigma, \tag{2.157}$$

whereas the second one, after using this identification, leads to the condition $\dot{H}_\sigma = 0$, which guarantees that the constraint (2.156) is preserved in time. Each of the equations of motion (2.149) leads to the field equation (2.110) again (where the second of these derivations, reminiscent of the derivation leading to equation (2.106) of the previous subsection, requires a little more work compared to the first one), and the combined effect of the equations of motion (2.150) is nothing but the field equation (2.97). Finally, both of the remaining two equations of motion (2.151), as well as their complex conjugates, give the Schrödinger-Kuchař equation (2.119), and its complex conjugate, respectively. As we elaborated in the previous subsection, given the relation (2.114) implied by the last of the five equations (2.153), equations (2.97) and (2.110) implied by equations (2.150) and (2.149) are equivalent to the Newton-Cartan field equations (2.98) and (2.118), respectively. Furthermore, the constraint (2.110) on the Cauchy data implied by the second of the five equations (2.153) is no constraint at all, because it is automatically satisfied by both of the Hamiltonian equations of motion (2.149); the constraint $u^\nu t_\nu = 1$ implied by the third of these equations, on the other hand, is trivially satisfied if we choose u^ν to be a unit timelike vector-field, as we have done.

The presence of the remaining constraints, of course, indicate that our phase-space is still too large and we have not isolated the 'true' dynamical degrees of freedom in our choice of configuration space. Note, however, that, among the remaining constraints, the set (2.96) implied by equations (2.152) consists of purely kinematical constraints in that its elements do not impose any relations among the genuinely dynamical degrees of freedom. Each of the constraint functions $h^{\mu\nu}t_\nu$, $\nabla_\mu h^{\nu\sigma}$, $\nabla_\mu t_\nu$, and $\partial_{[\mu} t_{\nu]}$ has vanishing variations with respect to the dynamical variables A_μ, u^ν, Ψ, and $\overline{\Psi}$ (recall that the Newton-Cartan connection appearing in the second and third of these functions is invariant under the changes of gauge variables A_μ and u^ν). This means that these are simply constant functions on the phase-space. In fact, we could have taken them as part of the background structure on the 'un-parameterized' manifold $\overset{\circ}{\mathcal{M}}$ before pulling them back with the fields by the map $\vartheta : \Sigma \to \overset{\circ}{\mathcal{M}}$. Consequently, if we ignore for the moment the kinematical constraints (2.156) and the linear constraints (2.154) and (2.155), what we have in hand is a Hamiltonian formulation of the classical Newton-Cartan-Schrödinger theory with the constraints (2.99) and (2.114) on the Cauchy data due to the last two of the constraint equations (2.153). If we use the field equation (2.110) implied by the Hamiltonian equations of motion (2.149) to substitute for $\nabla_\sigma \Theta^\sigma$ in the constraint (2.114), then it takes the form

$$\mathcal{C}(\kappa) \ := \ \frac{1}{\wp} \nabla_\sigma \Pi^\sigma \ + \ \Lambda_{\mathrm{N}} \ - \ (1-\kappa)\, 4\pi G\, \rho \ = \ \kappa \Lambda \,, \qquad (2.158)$$

where recall that κ is an arbitrary free parameter.

Now, we have seen that Newton-Cartan connection is invariant under internal gauge transformations $A_\mu \mapsto A_\mu + \partial_\mu f$ and boost transformations (2.127) over and above the diffeomorphisms $\phi \in \mathrm{Diff}(\mathcal{M})$ (cf. Eq. (2.44)). Physically this implies that the pairs $\{A_\mu, u^\nu\}$ and $\{(A_\mu + \partial_\mu f + \mathrm{w}_\mu - (u^\sigma \mathrm{w}_\sigma + \frac{1}{2} h^{\alpha\sigma} \mathrm{w}_\alpha \mathrm{w}_\sigma)$

t_μ), $(u^\nu + h^{\nu\sigma} \mathrm{w}_\sigma)\}$ of gravitational variables represent the same physical configuration of the gravitational field. On the other hand, as we noted in section 2.3, under the vertical transformations (2.44) the Schrödinger-Kuchař field transforms as $\Psi \mapsto \exp(i\frac{m}{\hbar} f)\, \Psi$, and, hence, the pair $\{\Psi, \overline{\Psi}\}$ representing matter fields is physically equivalent to the pair $\{(\exp(i\frac{m}{\hbar} f)\, \Psi), (\exp(-i\frac{m}{\hbar} f)\, \overline{\Psi})\}$. Therefore, we must replace the configuration space Z of our composite physical system by the space $\widetilde{Z}$ of *equivalence classes* of representatives of the physical fields, where all representatives of the fields are defined to be equivalent if they differ only by the interconnecting vertical gauge transformations (2.44). Now recall that in the canonical formalism, quite generally, the space of possible momenta of a physical system at a given configuration point z is the cotangent space $T_z^* Z$ of the configuration space at that point, and, consequently, the corresponding phase-space is a cotangent-bundle $T^* Z$ over the configuration space endowed with a natural presymplectic structure (cf. appendix); where, given a configuration manifold Z and a point z on that manifold, a cotangent vector $\mathcal{P}$ at the point is defined to be a real linear map, $\mathcal{P} : T_z Z \to \mathbb{R}$, from the tangent space at that point to the set of real numbers. For example, in the particular case under consideration, the momentum density Π^μ is a cotangent vector at the point A_μ of the configuration space which maps the tangent vector δA_μ at A_μ into $\mathbb{R}$ via $\delta A_\mu \to \int_{\Sigma_t} \Pi^\mu \, \delta A_\mu$. If we now take our configuration space to be the space $\widetilde{Z}$ of *equivalence classes* of representatives of the physical fields as defined above, then the cotangent space at a configuration point would be the space of real linear functions of variations of the physical fields which are independent of the vertical gauge transformations (2.44). Consequently, the momentum densities would be represented by those vector fields which leave the integral

$$\int_{\Sigma_t} d^3 x \left(\Pi^\mu \, \delta A_\mu \;+\; \overset{u}{\Pi}_\nu \, \delta u^\nu \;+\; \overline{\mathrm{P}} \, \delta \Psi \;+\; \mathrm{P} \, \delta \overline{\Psi} \;+\; \Pi_\sigma \, \delta \vartheta^\sigma \right) \qquad (2.159)$$

invariant under the transformations (2.44), where we freely use the convenient fact that $^{\parallel}\Pi^{\sigma}\,\delta_{\parallel}A_{\sigma} \equiv \Pi^{\sigma}\,\delta A_{\sigma}$. However, it is easy to see by direct substitutions for the expressions in the integrand (and using equation (2.11)) that this integral is left invariant under (2.44), or, effectively, under

$$
\begin{aligned}
A_{\mu} &\longrightarrow A_{\mu} + \nabla_{\mu}f \\
\Psi &\longrightarrow \exp(i\tfrac{m}{\hbar}f)\,\Psi \,,
\end{aligned} \tag{2.160}
$$

if and only if the momentum densities satisfy the condition

$$
\frac{1}{\wp}\nabla_{\sigma}\Pi^{\sigma} \,-\, 4\pi G\,\rho \,=\, 0 \,. \tag{2.161}
$$

Next, we wish to further restrict our configuration space and admit only those configuration variables which satisfy the constraint (2.99) implied by the fourth of the constraint equations (2.153). Now, as discussed in the previous subsection (cf. expressions (2.109) and (2.112)), this restriction amounts to an addition of a couple of 4-divergence terms in the integral (2.159), which, however, remains unaffected by them because of the vanishing of the boundary of a boundary theorem. Nevertheless, the ensuing condition (2.161) *is* modified under the transformations (2.160), and becomes

$$
\frac{1}{\wp}\nabla_{\sigma}\Pi^{\sigma} \,+\, \Lambda_{\mathrm{N}} \,-\, 4\pi G\,\rho \,=\, 0 \,, \tag{2.162}
$$

with the Λ_{N} term arising from the addition of the 4-divergence (2.109) in the integral (2.159).

But this is just the constraint (2.158) in the limit $\kappa \to 0$:

$$
\lim_{\kappa \to 0} \mathcal{C}(\kappa) \,=\, \frac{1}{\wp}\nabla_{\sigma}\Pi^{\sigma} \,+\, \Lambda_{\mathrm{N}} \,-\, 4\pi G\,\rho \,=\, 0 \tag{2.163}
$$

(recall that, as far as the field equations and their derivations are concerned, we are free to choose any desired value for κ without loss of generality). Further, in this harmless limit, not only the field equation (2.110) and the constraint equation (2.158) become identical, but also the Lagrangian density (2.92) becomes invariant under the full automorphism group (2.45). In summary, by eliminating the spurious gauge-arbitrariness in the configuration space by working rather with equivalence classes of field variables, we have been able to liberate our phase-space from the constraints (2.99) and (2.158). In fact we have arrived at the following set of results: in the light of equation (2.162) being a direct consequence of the invariance of the integral (2.159) under the vertical gauge transformations (2.44), the limit $\kappa \rightarrow 0$, (1), eliminates the unwanted constraint (2.158) on the Cauchy data, (2), dictates that the constituents of the configuration space of the system are automatically entire classes of representatives of the physical fields which, in addition, satisfy the constraint (2.99), (3), makes the Lagrangian density invariant under the full group $\mathcal{A}ut(B(\mathcal{M}))$, and, by virtue of (2), (3) and $^{\|}\Pi^{\sigma}\,_{\|}\dot{A}_{\sigma} \equiv \Pi^{\sigma}\,\dot{A}_{\sigma}$, (4), renders the Hamiltonian functional (2.147) manifestly invariant under the vertical gauge transformations (2.44). In the remaining of this paper the limit $\kappa \rightarrow 0$ will be understood to have been taken.

Let us now turn our attention to the remaining linear constraints (2.154) and (2.155). To eliminate these constraints, we further reduce our phase-space by defining a new, more appropriate set of canonical variables $\{v^{\alpha},\,\pi_{\alpha}\,;\,\psi,\,p\,;\,\vartheta^{\sigma},\,\Pi_{\sigma}\}$, where

$$v^{\alpha} := u^{\alpha} - h^{\alpha\sigma} A_{\sigma} = u^{\alpha} - h^{\alpha\sigma}\,_{\|}A_{\sigma}\,,$$

$$\pi_{\alpha} := {}_{\|}\overset{u}{\Pi}_{\alpha} = \overset{u}{\delta}_{\alpha}{}^{\mu}\,\overset{u}{\Pi}_{\mu} = \overset{u}{\Pi}_{\mu} = -\overset{u}{h}_{\alpha\sigma}\Pi^{\sigma}\,;$$

$$\psi := \frac{1}{\hbar\sqrt{8\pi G\,\wp}}\,(\,2\pi G\,\wp\,\hbar\,\overline{\Psi} + i\,\overline{P}),$$

and $$p := \frac{1}{\sqrt{8\pi G\,\wp}}\,(P + i\,2\pi G\,\wp\,\hbar\,\Psi),\tag{2.164}$$

so that equations (2.155) yield

$$\psi = \sqrt{2\pi G\,\wp}\,\overline{\Psi} \qquad \text{and} \qquad p = i\hbar\,\overline{\psi} \qquad (2.165)$$

(cf. Eqs. (2.134), (2.137), and (2.79)). In terms of these new canonical variables the Hamiltonian density (2.146) translates to be

$$\aleph^\sigma H_\sigma \equiv H := \pi_\sigma\,\dot{v}^\sigma + 2p\dot{\psi} + \Pi_\sigma\,\dot{\vartheta}^\sigma - \mathcal{L}(v^\alpha, \pi_\alpha; \psi, p; \vartheta^\sigma, \Pi_\sigma)\,, \quad (2.166)$$

which can be rewritten in the familiar form as

$$\aleph^\sigma H_\sigma \equiv H := \pi_\sigma\,\dot{v}^\sigma + p\dot{\psi} + \Pi_\sigma\,\dot{\vartheta}^\sigma - \mathcal{L}^{\text{c}}(v^\alpha, \pi_\alpha; \psi, p; \vartheta^\sigma, \Pi_\sigma)\,, \quad (2.167)$$

with

$$\mathcal{L}^{\text{c}} := \mathcal{L} - p\dot{\psi} \qquad (2.168)$$

being the 'correct' (or 'constraint-free') Lagrangian density. (Note that $\overline{\mathrm{P}}\,\dot{\Psi} + \mathrm{P}\,\dot{\overline{\Psi}}$ translates into $2\,p\,\dot{\psi}$ only upto a total time derivative, which, of course, does not affect the action.) The 'correct' Hamiltonian form of the parameterized action functional now reads

$$\mathcal{I}_{\mathbb{H}}^{\text{c}} = \int_{\mathbb{R}} dt \int_{\Sigma_t} d^3x\,[\pi_\sigma\,\dot{v}^\sigma + p\dot{\psi} + \Pi_\sigma\,\dot{\vartheta}^\sigma - \aleph^\sigma H_\sigma]\,, \qquad (2.169)$$

with the Hamiltonian equations of motion

$$\frac{\delta H}{\delta\,\Pi_\alpha} = +u^\sigma\nabla_\sigma\,\vartheta^\alpha\,, \qquad\qquad \frac{\delta H}{\delta\,\vartheta^\alpha} = -u^\sigma\nabla_\sigma\,\Pi_\alpha\,, \qquad (2.170a)$$

$$\frac{\delta H}{\delta\,\pi_\alpha} = +u^\sigma\nabla_\sigma\,v^\alpha\,, \qquad\qquad \frac{\delta H}{\delta\,v^\alpha} = -u^\sigma\nabla_\sigma\,\pi_\alpha\,, \qquad (2.170b)$$

$$\text{and} \qquad \frac{\delta H}{\delta\,p} = +u^\sigma\nabla_\sigma\,\psi\,, \qquad\qquad \frac{\delta H}{\delta\,\psi} = -u^\sigma\nabla_\sigma\,p\,. \qquad (2.170c)$$

$$(2.170)$$

It is easy to check that this set of equations of motion is equivalent to the previous set with the *same* Hamiltonian density, but, of course, expressed in terms of the old set of somewhat redundant canonical variables. The Hamiltonian density (2.167) clearly suggests the 'correct' Lagrangian form of the action functional,

$$\mathcal{I}^{c} = \int_{\mathcal{O}} d^4x \; \mathcal{L}^{c}(v^{\sigma}, \nabla_{\mu}v^{\sigma}, \nabla_{\mu}\nabla_{\nu}v^{\sigma}, \psi, \partial_{\mu}\psi \, ; \, {}^{(s)}y) \,, \qquad (2.171)$$

which, *a priori*, might be taken to indicate that we would have been better off taking from the beginning the alternative form

$$\mathcal{L}^{c}_{\Psi} = +\wp \, 4\pi G \left\{ \frac{\hbar^2}{2m} h^{\alpha\beta} \partial_{\alpha}\Psi \partial_{\beta}\overline{\Psi} - i\frac{\hbar}{2} u^{\alpha}(\overline{\Psi}\partial_{\alpha}\Psi) \right\} \qquad (2.172)$$

of the component $\mathcal{L}_{\Psi}$ in the action integral (2.91). However, this form of the Lagrangian density is not Hermitian and breaks the symmetry which naturally exists between Ψ and $\overline{\Psi}$. Moreover, it contains an incorrect factor of $\frac{1}{2}$ in the second term. Nevertheless, in the end, it is this form (2.172) which turns out to be the most appropriate one as far as the new set of canonical variables are concerned.

Finally, what about the last remaining constraint — the diffeomorphism constraint (2.156) due to the parameterization process — which we have left on hold so far? It turns out that this constraint (2.156) can also be readily untangled and, unlike the parallel case in the Hamiltonian formulation of the 'already parameterized' Einstein's theory of gravity, here the process of 'deparameterization' [38] can be easily carried out. This is because in our case the kinematical momentum densities Π_{σ} conjugate to the supplementary embedding variables ϑ^{σ} appear *linearly* in the expression of the constraint, neatly segregating themselves from the set of genuinely dynamical variables. Consequently, all one has to do to recover the 'true' Hamiltonian form of the physical action is to solve the constraint for these momentum densities and get $\Pi_{\sigma} = -\overset{\circ}{H}_{\sigma}$, substitute this result into the Hamiltonian action functional (2.169), and then impose the relation (2.157) (obtained from one of the equations of motion) on the resulting expression to complete the deparameterization.

Thus, we have been able to eliminate *all* redundancies from the configuration space by redefining it to consist only true dynamical degrees of freedom, and, as a result, succeeded in constructing a meaningful (constraint-free) phase-space for our Newton-Cartan-Schrödinger system. Let us now recapitulate the main features of this phase-space. First, and foremost, as there are no topological complications ($\Sigma_t \cong \mathbb{R}^3$), the phase-space is naturally a *cotangent-bundle* $T^*\tilde{Z}$ over the infinite-dimensional complex configuration space $\tilde{Z}$ of equivalence classes of fields $\{\tilde{v}^\sigma(x);\ \tilde{\psi}(x)\}$ evaluated on the 3-submanifold Σ of $\mathcal{M}$:

$$\tilde{Z} := \{v^\sigma; \psi \mid t_\sigma v^\sigma = 1,\ \nabla_{[\mu}\nabla_{\nu]}v^\sigma = 0,\ v^\sigma \sim v^\sigma - h^{\sigma\mu}\nabla_\mu f\ ;\ \psi \sim \exp(-i\tfrac{m}{\hbar}f)\psi\}$$

(2.173)

where $\nabla_{[\mu}\nabla_{\nu]}v^\sigma = 0$ is equivalent to the condition (2.99), and $v^\sigma \sim v^\sigma - h^{\sigma\mu}\nabla_\mu f$ results form the equivalence (2.160). What is more, this cotangent-bundle possesses a well-defined, *non-degenerate* symplectic structure, which is just its natural symplectic structure

$$\tilde{\omega} = \int_{\Sigma_t} d^3x\, [\, d_z\tilde{\pi}_\mu \wedge d_z\tilde{v}^\mu + d_z\tilde{p} \wedge d_z\tilde{\psi}\,] \qquad (2.174)$$

with $\tilde{v}^\alpha$, $\tilde{\pi}_\alpha$, $\tilde{\psi}$, and $\tilde{p}$ representing entire gauge-equivalence classes of fields as discussed above. Note that, although $\tilde{Z}$ has a natural vector-space structure over the *complex* field $\mathbb{C}$, the 2-form $\tilde{\omega}$ is *real-valued* since $p = i \times \overline{\psi}$.

As it stands, however, it is not immediately clear whether this description of symplectic structure is a generally-covariant one. In fact it appears not to be a covariant one on two counts. First of all, in the present subsection we have been working strictly within a non-covariant 3+1 decomposition of spacetime. Secondly, the symplectic structure (2.174) we have arrived at in the end corresponds to the *deparameterized* action — i.e., an action effectively corresponding to the original unparameterized theory on a fixed background $(\mathcal{M};\ h,\ \tau)$. As we shall see in the next subsection, however, the expression (2.174) in fact provides a truly generally-covariant description of the phase-space, provided it is viewed more appropriately.

2.4.3 Manifestly covariant description of the canonical formulation

In the previous subsection we have constructed a canonical phase-space for the Newton-Cartan-Schrödinger system by working within a 3+1 decomposition of spacetime into 3-spaces at instants of time. Such a blatantly non-covariant breakup of spacetime is admittedly somewhat natural for a Newtonian theory, but it undermines the efforts of Cartan and followers to give a full spacetime-covariant meaning to Newtonian gravity by violating manifest covariance of even the old-fashioned Galilean kind. Therefore, at least for the sake of Cartan's legacy if not for anything else, it is desirable to seek a manifestly covariant version of the canonical phase-space constructed in the previous subsection. Fortunately, it has long been recognized — dating all the way back to Lagrange [51] — that the phase-space of a physical system is better viewed as the space of entire dynamical histories of the system, without reference to a particular instant of time. With this view of phase-space, the core of the canonical formalism can be developed in a manner that manifestly preserves all relevant symmetries of a given classical system [44, 45, 49–51]. Since this 'covariant phase-space formalism' is relatively less-popular, and since we shall be applying it to our Newton-Cartan-Schrödinger system in the present subsection as well as in the subsection 2.5.2 below when we quantize the system, we have briefly reviewed its main concepts in the appendix A below setting our notational conventions. We urge the reader not familiar with these concepts to carefully study the appendix before reading any further in order to appreciate the elegance of the underlying ideas used in what follows.

The essence of the covariant phase-space — the space $\mathcal{Z}$ of solutions of the equations of motion of a theory — is most succinctly encapsulated in a closed, non-degenerate symplectic 2-form $\tilde{\omega}$ defined on the quotient space $\tilde{\mathcal{Z}} := \mathcal{Z}/\mathcal{K}$, where $\mathcal{K}$ is the characteristic distribution defined by equation (2.228) of the appendix. As a first step towards evaluating $\tilde{\omega}$ for our Newton-Cartan-Schrödinger system, we work out the presymplectic potential current density (2.219) for the generally-covariant action functional (2.171),

$$\mathcal{J}_{\mathcal{Q}}^{\mu} = \mathcal{J}_{v}^{\mu} + \mathcal{J}_{\psi}^{\mu} + \mathcal{J}_{(s)y}^{\mu} , \tag{2.175}$$

giving

$$\mathcal{J}_{\mathcal{Q}}^{\alpha} t_{\alpha} = \pi_{\mu} \delta v^{\mu} + p \, \delta \psi + {}^{(s)}\Pi_{\sigma} \delta \, {}^{(s)}y^{\sigma} . \tag{2.176}$$

If we assume that $\mathcal{J}_{\mathcal{Q}}^{\alpha} \to 0$ at spatial infinity (or work with a compact Σ_t), then the corresponding *presymplectic* 2-form ω is given by the equation (2.214), with $\omega^{\mu} = d_z \mathcal{J}_{\mathcal{Q}}^{\alpha}$ (cf. Eq. (2.226)):

$$\omega = \int_{\Sigma_t} d^3x \, [\omega^{\mu} t_{\mu}] = \int_{\Sigma_t} d^3x \, [d_z \pi_{\mu} \wedge d_z v^{\mu} + d_z p \wedge d_z \psi + d_z {}^{(s)}\Pi_{\sigma} \wedge d_z {}^{(s)}y^{\sigma}]. \tag{2.177}$$

Now, in accordance with the discussion around equation (2.228) of the appendix, this presymplectic structure has a degenerate direction for each infinitesimal gauge transformation of the theory stemming from the action of the automorphism group $\mathcal{A}ut(B(\mathcal{M}))$. Consequently, we seek the non-degenerate projection $\tilde{\omega}$ of ω on the physically relevant *reduced phase-space* $\tilde{\mathcal{Z}} = \mathcal{Z}/\mathcal{A}ut(B(\mathcal{M}))$, which is nothing but the space of orbits of the group $\mathcal{A}ut(B(\mathcal{M}))$ in the solution-space $\mathcal{Z}$. Fortunately, the complete automorphism group has the structure of a semidirect product: $\mathcal{A}ut(B(\mathcal{M})) = \mathcal{V}(B(\mathcal{M})) \circledS \mathrm{Diff}(\mathcal{M})$, where $\mathcal{V}(B(\mathcal{M}))$ is the group of vertical gauge transformations given by (2.46). This makes it possible to discuss effects of the two subgroups successively, since the isomorphism

$$\mathcal{Z}/\mathcal{A}ut(B(\mathcal{M})) \cong [\mathcal{Z}/\mathcal{V}(B(\mathcal{M}))]/\mathrm{Diff}(\mathcal{M}) \tag{2.178}$$

holds. First note that, in analogy with the discussion in the previous subsection, the quotient space $\mathcal{Z}/\mathcal{V}(B(\mathcal{M}))$ is the space of solutions modulo vertical gauge directions in which the integral $\int_{\Sigma_t} d^3x\, \mathcal{J}_\mathcal{Q}^\alpha t_\alpha$ is rendered gauge-invariant (cf. Eq. (2.159)). In terms of the projection map $\mathbb{P} : \mathcal{Z} \to \tilde{\mathcal{Z}}$, this implies

$$\mathbb{P}(v) = \tilde{v}, \qquad \mathbb{P}(\pi) = \tilde{\pi}, \qquad \mathbb{P}(\psi) = \tilde{\psi}, \qquad \text{and} \qquad \mathbb{P}(p) = \tilde{p}. \qquad (2.179)$$

Next, a globally valid gauge representing the moduli space $[\mathcal{Z}/\mathcal{V}(B(\mathcal{M}))]/\text{Diff}$ $(\mathcal{M})$ can be easily constructed by simply taking ${}^{(s)}y : \mathcal{M} \to \overset{\circ}{\mathcal{M}}$ to be a fixed diffeomorphism ${}^{(o)}y$ (e.g., ${}^{(o)}y = identity$) and then specifying the values of the dynamical variables on Σ_t with respect to this choice. This is possible because, as in a parameterized scalar field theory in Minkowski spacetime [44, 46], any arbitrary variation of the diffeomorphism ${}^{(s)}y$ sweeps out the entire degeneracy submanifold of ω, allowing us to uniquely characterize each degeneracy sub-manifold by a fixed diffeomorphism ${}^{(o)}y$. But this immediately renders $\mathcal{J}^\mu_{{}^{(o)}y} = 0$ (cf. Eq. (2.219)). Therefore, the net result of projecting the presymplectic 2-form ω on the reduced phase-space $\tilde{\mathcal{Z}} = \mathcal{Z}/\mathcal{A}ut(B(\mathcal{M}))$ by the projection map $\mathbb{P} : \mathcal{Z} \to \tilde{\mathcal{Z}}$ is the desired non-degenerate 2-form

$$\tilde{\omega} = \int_{\Sigma_t} d^3x\, [\, d_z \tilde{\pi}_\mu \wedge d_z \tilde{v}^\mu + d_z \tilde{p} \wedge d_z \tilde{\psi}\,]. \qquad (2.180)$$

It is easy to see that ω in expression (2.177) is a unique pull-back $\mathbb{P}^*(\tilde{\omega})$ of $\tilde{\omega}$ form $\tilde{\mathcal{Z}}$ to $\mathcal{Z}$. In other words, we have

$$\underset{A}{Y} \lrcorner\, \tilde{\omega} = 0 \qquad (2.181)$$

(cf. Eq. (2.230)), where $\underset{A}{Y}$ are the gauge directions corresponding to the complete automorphism group $\mathcal{A}ut(B(\mathcal{M}))$. Again in close analogy with the covariant phase-space description of a parameterized scalar field theory in Minkowski spacetime [44, 46], this non-degenerate symplectic 2-form $\tilde{\omega}$ describes precisely the phase-space of the original unparameterized theory in the fixed, non-dynamical spacetime $(\mathcal{M}; h^{\mu\nu}, t_\mu)$. Thus, the reduction procedure in the present case has the effect of 'deparameterizing' the parameterized theory, just as in the case of a scalar field theory. However, unlike in the previous subsection, here our description is manifestly covariant. Further, the resulting *constraint submanifold* [44] $\widetilde{\mathcal{Z}}$ of the dynamically possible states of the system permeates *all* of the space $\widetilde{Z}$ of kinematically possible states because the constraints have been eliminated (cf. Eq. (2.231)). In fact, as comparison of equations (2.174) and (2.180) immediately suggests, the covariant phase-space $\widetilde{\mathcal{Z}}$ of the present subsection is symplectically diffeomorphic to the constraint-free canonical phase-space $T^*\widetilde{Z}$ of the previous subsection:

$$\widetilde{\mathcal{Z}} \simeq T^*\widetilde{Z}. \tag{2.182}$$

It is worth recalling here that the symplectic structure $\widetilde{\omega}$ is independent of the choice of a hypersurface Σ_t, and as such, in particular, it is Milne-invariant (cf. Eq. (2.54)). More generally, $\widetilde{\omega}$ is rather trivially invariant under the action of $\mathrm{Diff}(\mathcal{M})$ on $\mathcal{Z}$ since all ingredients in its expression above transform homogeneously, like tensors. In short, we have been able to successfully crystallize the diffeomorphism-invariant canonical essence of Newton-Cartan-Schrödinger system in the expression (2.180).

2.5 Quantization of the Newton-Cartan-Schrödinger system

So far we have considered only classical, external gravitational field. However, as noted before, in Newton-Cartan theory the connection-field is a dynamical object: it is not just a part of the immutable background structure, but depends crucially on the distribution of matter-sources via the field equation

$R_{\mu\nu} = 4\pi G\, M_{\mu\nu}$. Since this equation dictates the coupling of spacetime curvature to *quantum mechanically* treated matter, the Newton-Cartan connection cannot have an *a priori* definite value and must itself be treated in a quantum mechanical fashion. Thus, a consistent account of physical phenomena even at a Galilean-relativistic level *necessitates* the construction of a quantum theory of gravity in which the superposition principle holds not only for the states of matter, but also for the states of the Newton-Cartan connection-field. In what follows we construct such a generally-covariant, Milne-relativistic quantum theory of gravity in which quantized Schrödinger particles produce the *quantized* Newton-Cartan connection-field through which they interact.

2.5.1 Covariant phase-space quantization

Having successfully identified the constraint-free phase-space for our classical Newton-Cartan-Schrödinger system in the previous section, the desired quantum theory can be easily constructed using a manifestly covariant approach to the usual canonical quantization method. An accessible reference on the general procedure is [54]. Recall that classical observables, say O, are maps from the phase-space $\widetilde{\mathcal{Z}}$ to $\mathbb{R}$, and the non-degenerate symplectic 2-form $\widetilde{\omega}$ on $\widetilde{\mathcal{Z}}$ determines a set of Poisson brackets inducing a Lie-algebra structure on the space of these observables. To quantized such a system, we are supposed to replace the classical observables with operators and Poisson brackets with commutators providing a corresponding algebraic structure on the space of these operators. More precisely, we are to seek a correspondence map, $\widehat{}: \mathsf{O} \mapsto \widehat{\mathsf{O}}$, and look for an irreducible representation of the Lie-algebra of classical observables as an algebra of operators acting on elements of some separable Hilbert space $\mathcal{H}$. It is well-known, however, that, as stated, this programme cannot be carried out in general. As early as in 1951 van Hove demonstrated that, for theories in which the position and momentum operators are represented in the standard manner, no such correspondence map can provide an irreducible representation of the full Poisson algebra of classical observables [55]. Fortunately, in our case the phase-space is naturally isomorphic to a cotangent-bundle, $\widetilde{\mathcal{Z}} \cong T^*\widetilde{Z}$, for which the van Hove obstruction is neutralised. Consequently, for us, it will turn out to be possible to choose a Hilbert space $\mathcal{H}$ and a correspondence map $\hat{\ } : O \mapsto \hat{O}$ such that the non-vanishing commutators satisfy[8]

$$[\,\widehat{\widetilde{v}}^{\,\mu}(x),\,\widehat{\widetilde{\pi}}_\nu(x')\,] \;=\; i\hbar\,\widehat{\mathbb{1}}\,\delta^\mu_{\ \nu}\,\delta(\vec{x}-\vec{x}') \qquad \text{and} \qquad [\,\widehat{\widetilde{\psi}}(x),\,\widehat{\widetilde{\psi}}^{\,\dagger}(x')\,] \;=\; \widehat{\mathbb{1}}\,\delta(\vec{x}-\vec{x}') \tag{2.183}$$

at equal-times (cf. Eq. (2.81)).

(8. For definiteness and simplicity, we shall only follow the bosonic case here. In this Galilean-relativistic context, where there is no connection between spin and statistics, a parallel discussion with a fermionic field is relatively straightforward (see subsection 2.4.7 of Ref.54 for a fermionic treatment in the relativistic case.))

The appearance of the Dirac delta-'function' here necessitates that $\widehat{\widetilde{v}}^{\,\mu}$, $\widehat{\widetilde{\pi}}_\nu$, $\widehat{\widetilde{\psi}}$, and $\widehat{\widetilde{\psi}}^{\,\dagger}$ must all be viewed as operator-valued distributions, and indicates that only the fields smeared with appropriate test-functions have physical meaning as observables in accordance with the well-known analysis of measurements of field-observables developed by Bohr and Rosenfeld [57]. As they stand, however, these commutation relations are clearly not expressed in a manifestly covariant manner. This deficiency can be quite easily removed in our case, partly by exploiting the natural vector-space structure of the phase-space $\widetilde{\mathcal{Z}}$ noted in the previous section. On account of this vector-space structure of $\widetilde{\mathcal{Z}}$, we may identify the tangent space at any point $z \in \widetilde{\mathcal{Z}}$ with $\widetilde{\mathcal{Z}}$ itself. Furthermore, as discussed in the appendix, under this identification the symplectic form $\widetilde{\omega}$ becomes an antisymmetric bilinear function, $\widetilde{\omega} : \widetilde{\mathcal{Z}} \times \widetilde{\mathcal{Z}} \to \mathbb{R}$, on the resultant symplectic vector-space $\tilde{\mathcal{Z}}$. Thus, we may rewrite the commutation relations (2.183) in terms of the operators $\hat{\tilde{\omega}}(\mathcal{Q}, \cdot)$ corresponding to the functions $\tilde{\omega}(\mathcal{Q}, \cdot)$ as the single commutator

$$[\hat{\tilde{\omega}}(\mathcal{Q}_1, \cdot), \hat{\tilde{\omega}}(\mathcal{Q}_2, \cdot)] = -i\hbar\, \hat{\mathbb{1}}\, \tilde{\omega}(\mathcal{Q}_1, \mathcal{Q}_2)\,, \qquad (2.184)$$

where $\mathcal{Q}, \mathcal{Q}_1, \mathcal{Q}_2 \in \tilde{\mathcal{Z}}$ (see [54] for further details on such translations). Since the self-adjoint operators appearing in this expression are unbounded, and, hence, only densely defined, it is convenient to work with the equivalent but better behaved *Weyl relations*,

$$\widehat{W}^\dagger(\mathcal{Q}) = \widehat{W}(-\mathcal{Q})$$
$$\text{and} \quad \widehat{W}(\mathcal{Q}_1)\,\widehat{W}(\mathcal{Q}_2) = \exp[\tfrac{i}{2}\hbar\, \tilde{\omega}(\mathcal{Q}_1, \mathcal{Q}_2)]\,\widehat{W}(\mathcal{Q}_1 + \mathcal{Q}_2)\,, \qquad (2.185)$$

where

$$W(\mathcal{Q}) := \exp[i\,\tilde{\omega}(\mathcal{Q}, \cdot)] \qquad (2.186)$$

is unitary and varies with $\mathcal{Q}$ in the 'strong operator topology' [56].

2.5.2 The GNS-construction and the choice of a vacuum state

It is well-known that the set, $\mathcal{B}(\mathcal{H})$, of all bounded linear maps on $\mathcal{H}$ has the natural structure of a C*-algebra with the '$*$-operation' corresponding to taking adjoints. The subalgebra $\mathcal{A}$ of $\mathcal{B}(\mathcal{H})$ generated by $\{\widehat{W}(\mathcal{Q})\,|\,\mathcal{Q} \in \tilde{\mathcal{Z}}\}$ satisfying the above relations is called the *Weyl algebra* over the symplectic vector-space $\tilde{\mathcal{Z}}$. Each normalized, positive algebraic state $\zeta : \mathcal{A} \to \mathbb{C}$ over the Weyl algebra $\mathcal{A}$ — viewed as an abstract C*-algebra — determines a Hilbert space $\mathcal{H}_\zeta$ and a representation $\mathcal{R}_\zeta : \mathcal{A} \to \mathcal{B}(\mathcal{H}_\zeta)$ of $\mathcal{A}$ by bounded linear operators acting in $\mathcal{H}_\zeta$, and thereby defines an Hermitian scalar product on $\mathcal{A}$ by

$$\zeta(a^*b) := \langle ab \rangle \qquad \forall\, a, b \in \mathcal{A}\,. \qquad (2.187)$$

Conversely, each choice of a measure $\mu : \mathcal{H}_\zeta \times \mathcal{H}_\zeta \to \mathbb{R}$ generates a state ζ on the algebra $\mathcal{A}$. Consequently, the positivity and normalization conditions on ζ can be expressed as

$$\zeta(a^*a) \geq 0 \quad \forall\, a \in \mathcal{A}, \quad \text{and} \quad \zeta(\mathbb{1}) = 1, \quad\quad (2.188)$$

where $\mathbb{1}$ denotes the identity element of $\mathcal{A}$. Such a construction of a representation of the Weyl algebra $\mathcal{A}$ as an algebra of operators on a Hilbert space $\mathcal{H}_\zeta$ is nothing but the celebrated GNS-construction [56], which, given a *cyclic* vector $|\xi^o\rangle \in \mathcal{H}_\zeta$, guarantees the existence of a representation $\mathcal{R}_\zeta : \mathcal{A} \to \mathcal{B}(\mathcal{H}_\zeta)$ such that

$$\zeta_{\xi^o}(a) = \langle \xi^o\, \mathcal{R}_\zeta(a)\xi^o \rangle \quad \forall\, a \in \mathcal{A}. \quad\quad (2.189)$$

The vector ξ^o is called a cyclic vector because the set $\{\, \mathcal{R}_\zeta(a)|\xi^o\rangle,\, |\, a \in \mathcal{A}\,\}$ constitutes a *dense* subspace of $\mathcal{H}_\zeta$. Upto unitary equivalence, the triplet $(\mathcal{H}_\zeta,\, \mathcal{R}_\zeta,\, |\xi^o\rangle)$ is uniquely determined by these properties, with the cyclic vector $|\xi^o\rangle, \in \mathcal{H}_\zeta$ corresponding to the identity element of $\mathcal{A}$.

Thus, an algebraic state ζ in $\mathcal{A}$ can be easily represented in the Hilbert space $\mathcal{H}_\zeta$ by the state-vector $|\xi^o\rangle$. Note, however, that the states over an abstract C*-algebra like $\mathcal{A}$ come in families: any vector $|\xi^i\rangle$ from a collection $\{\, |\xi^i\rangle\, |\, |\xi^i\rangle \in \mathcal{H}_\zeta\,\}$ represents the family of algebraic state ζ by

$$\zeta_{\xi^i}(a) = \langle \xi^i\, |\mathcal{R}_\zeta(a)|\xi^i\rangle, . \quad\quad (2.190)$$

(More generally, we may consider the states $\zeta_\mathcal{D}(a) = Tr[\mathcal{D}\,\mathcal{R}_\zeta(a)]$, with $\mathcal{D}$ being a positive trace-class operator in $\mathcal{B}(\mathcal{H}_\zeta)$). This is due to the fact that $\zeta_{\xi^i}(a)$ can be approximated as closely as desired by $\zeta_{\xi^o}(b^*ab)$, for any $b \in \mathcal{A}$, since, thanks to the cyclicity of $|\xi^o\rangle$, $|\xi^i\rangle$ can be approximated as closely as desired by $\mathcal{R}_\zeta(b)|\xi^o\rangle$:

$$\zeta_{\xi^i}(a) = \langle \xi^i\, \mathcal{R}_\zeta(a)\rangle\xi^i \approx \langle \xi^o\, \mathcal{R}_\zeta^*(b)\mathcal{R}_\zeta(a)\mathcal{R}_\zeta(b)|\xi^o\rangle = \zeta_{\xi^o}(b^*ab). \quad\quad (2.191)$$

Consequently, a GNS-representation crucially depends on the generating state ζ of the system. In particular, with changes in the state ζ, the measure μ — with respect to which the inner-product of the Hilbert space $\mathcal{H}_\zeta$ has been defined — changes. But this is a bad news: measures with different null-sets would in general lead to operators with different norms in the corresponding representations, and the kernel (i.e., the set of those operators which are mapped onto zero) of the representations will correspondingly differ. Consequently, such representations (uncountably many of them) will be *unitarily inequivalent* in general since the states which generate them could fail to determine quasi-equivalent measures — i.e., measures with the same null-sets. This, of course, is a well-known problem for quantum systems with infinitely many degrees of freedom, since the Stone-von Neumann uniqueness theorem is inapplicable for such systems [54].

Fortunately, if we are willing to let go a bit of the mathematical elegance maintained so far and find our way back to physics, there are two strategies at our disposal to tackle this problem. The first obvious strategy is to select a privileged cyclic state-vector using some physical criterion (some rule external to the quantum theory proper), and thereby obtain an equivalence class of representations which contains this distinguished state. The standard choice in Minkowskian quantum field theories is, of course, the *vacuum* state, $|\xi^o\rangle \equiv |0\rangle$, which is required to remain invariant under the action of the Poincaré group — the isometry group of the flat Minkowski spacetime. Then, through GNS-construction, the vacuum expectation values of all operators provide the Hilbert space equipped with a natural inner-product. This choice of a GNS-representation is then simply the Fock-representation. In our case it is most natural to use the Milne group defined by

$$0 = \mathcal{L}_x h^{\mu\nu} = \mathcal{L}_x t_\mu = \mathcal{L}_x \Gamma^{\alpha\nu}_\mu \qquad (2.192)$$

(cf. Eq. (2.56)) in place of the Poincaré group in the above procedure, and take the Milne-invariant no-particle state as our privileged state. The Hilbert space we thereby obtain is a Milne-relativistic Fock-representation of the Weyl algebra for our Newton-Cartan-Schrödinger system.

The second strategy to deal with the problem of inequivalent representations is to follow the general 'operational' philosophy historically motivated by a result of Fell [58], and adopt his criterion of 'physical equivalence' as opposed to the strict mathematical equivalence. This too is a well-known strategy, a good accessible account of which can be found in [54] (see, especially, his Theorem 4.5.2). The idea is to first acknowledge the finiteness of accuracy and number of possible realistic measurements, and then realize that, due to these limitations, it is physically impossible to distinguish between the 'inequivalent' representations of the C*-algebra $\mathcal{A}$. Consequently, a choice form the myriad of inequivalent representations is physically irrelevant, and the choice of Milne-invariant Fock-representation made above is as good as any.

2.5.3 The Hamiltonian operator: in general and in an inertial frame

Having settled the problem of inequivalent representations does not, of course, guarantee that every important observable needed to unambiguously define a quantum theory is contained in the algebra $\mathcal{A}$. Therefore, let us verify that at least the most important observables of our theory are well-defined. We begin with one of the simplest operator: the covariant mass-density operator

$$\widehat{M_{\mu\nu}} := m\,\widehat{\widetilde{\psi}}^{\dagger}\,\widehat{\widetilde{\psi}}\,t_{\mu\nu} \tag{2.193}$$

(cf. Eq. (2.85)). We have been careful to choose an appropriate ordering in defining this operator, and, as a result, it is manifestly well-defined. Similarly, the operator corresponding to the Riemann tensor (2.26) is also well-defined provided the operators corresponding to the gravitational field appearing in its polynomial expression are properly normal-ordered. On the other hand, the all important Hamiltonian-density operator

$$\widehat{H} \;:=\; H(\widehat{\widetilde{v}}^{\,\alpha},\,\widehat{\widetilde{\pi}}_{\alpha}\,;\,\widehat{\widetilde{\psi}},\,\widehat{\widetilde{\psi}}^{\dagger}) \tag{2.194}$$

in our interacting theory involves both matter and gravitational field variables. Hence, *a priori*, one might fear existence of potential operator ordering ambiguities in its expression. However, recall that all of the matter variables commute with all of the gravitational variables, removing any danger of intractable ordering ambiguity. Consequently, the Hamiltonian operator

$$\widehat{\mathbb{H}} \;:=\; \int_{\Sigma_t} :\widehat{H}:\; d^3x \tag{2.195}$$

is also well-defined, where matter and gravitational operators in $\widehat{H}$ are taken to be independently normal-ordered.

The vanishing commutation between matter and gravitational variables imply, in particular, that there is no gravitational self-interaction in this linear, Newtonian quantum gravity; i.e., the Schrödinger-Fock particles of the system do not gravitationally self-interact, though they interact among themselves. This result becomes most conspicuous from the perspective of an observer confined to a *local* inertial (i.e., Galilean) frame of reference equipped with a Cartesian coordinate system. In a local inertial frame the inertial and gravitational parts of the Newton-Cartan connection-field may be unambiguously (but, of course, non-covariantly) separated as in the equation (2.24) above, and a linear coordinate system may be introduced such that the two metric fields assume their canonical forms: $h = \delta^{ab}\partial_a \otimes \partial_b$ and $\tau = dt$. Furthermore, the connection-field in such a frame corresponds to a gauge-choice $u = \frac{\partial}{\partial t}$ and $A = -\Phi\tau$ in equation

(2.30), with Φ viewed as the usual Newtonian gravitational potential. With a further gauge-choice of $\chi = -\frac{1}{2}\Phi$, and setting the rest of the multiplier fields to zero (without any loss of information, of course), the action functional (2.91) in the inertial frame becomes

$$\mathcal{I} = \int dt \int d\vec{x}\, [\frac{1}{8\pi G}\, \Phi\, \Delta\, \Phi + \frac{\hbar^2}{2m}\, \delta^{ab}\, \partial_a \Psi\, \partial_b \Psi + i\frac{\hbar}{2}(\Psi\, \partial_t \overline{\Psi} - \overline{\Psi}\, \partial_t \Psi) - m\, \Psi\overline{\Psi}\, \Phi],$$
(2.196)

where we recall that $\kappa = 0$, and for simplicity we also take $\Lambda_{\rm o} = 0$. Extremization of this functional with respect to variations of the scalar potential Φ immediately yields the Newton-Poisson equation

$$\Delta\Phi \;=\; \frac{4\pi G}{\langle \psi\psi\rangle}\, m\, \psi\, \overline{\psi}\,,$$
(2.197)

where $\psi := \sqrt{2\pi G \wp}\; \overline{\Psi}$, and we set $\langle \psi\psi\rangle := \int d\vec{x}\, \overline{\psi}\, \psi = 1$. On the other hand, extremization of the action with respect to variations of the matter field Ψ leads to the familiar Schrödinger equation in the presence of an external gravitational field:

$$i\hbar \frac{\partial}{\partial t}\, \psi \;=\; [-\frac{\hbar^2}{2m}\Delta + m\, \Phi]\psi\,.$$
(2.198)

The last two equations may be interpreted as describing a *single* Galilean-relativistic particle gravitationally interacting with its own Newtonian field. As such, the coupled equations (2.197) and (2.198) constitute a *nonlinear* system, which can be easily seen as such by first formally solving equation (2.197) for the gravitational potential giving

$$\Phi(\vec{x}) \;=\; -\, Gm \int d\vec{x}'\, \frac{\overline{\psi}(\vec{x}')\, \psi(\vec{x}')}{|\vec{x}' - \vec{x}|}\,,$$
(2.199)

and then — by substituting this solution into equation (2.198) — obtaining the nonlinear integro-differential equation

$$i\hbar \frac{\partial}{\partial t}\, \psi(\vec{x}, t) \;=\; -\frac{\hbar^2}{2m}\Delta\, \psi(\vec{x}, t) \,-\, Gm^2 \int d\vec{x}'\, \frac{\overline{\psi}(\vec{x}', t)\, \psi(\vec{x}', t)}{|\vec{x}' - \vec{x}|}\, \psi(\vec{x}, t)\,.$$
(2.200)

However, when ψ is promoted to a 'second-quantized' field operator $\widehat{\psi}$ satisfying

$$[\widehat{\psi}(\vec{x}),\, \widehat{\psi}^{\dagger}(\vec{x}')] \;=\; \widehat{\mathbb{1}}\, \delta(\vec{x} - \vec{x}')\,, \qquad (2.201)$$

this equation describes a system of many identical particles in the Heisenberg picture, with $\widehat{\psi}$ acting as an annihilation operator in the corresponding Fock space, analogous to the covariantly described 'free' system discussed in the subsection 2.3.2 above. In particular, the normal-ordered Hamiltonian operator for the system now reads

$$
\begin{aligned}
\widehat{\mathbb{H}} \;&=\; \widehat{\mathbb{H}}_{\mathrm{o}} + \widehat{\mathbb{H}}_{\mathrm{I}}\\
\text{with} \qquad \widehat{\mathbb{H}}_{\mathrm{o}} \;&:=\; \int d\vec{x}\; \widehat{\psi}^{\dagger}(\vec{x})[-\frac{\hbar^2}{2m}\Delta]\widehat{\psi}(\vec{x})\,,\\
\text{and} \qquad \widehat{\mathbb{H}}_{\mathrm{I}} \;&:=\; -\frac{1}{2}\,Gm^2 \int d\vec{x}\int d\vec{x}'\; \frac{\widehat{\psi}^{\dagger}(\vec{x}')\,\widehat{\psi}^{\dagger}(\vec{x})\,\widehat{\psi}(\vec{x})\,\widehat{\psi}(\vec{x}')}{|\vec{x}' - \vec{x}|}\,,
\end{aligned} \qquad (2.202)
$$

which, upon substitution into the Heisenberg equation of motion

$$i\hbar\frac{\partial}{\partial t}\,\widehat{\psi}(\vec{x},\,t) \;=\; [\widehat{\psi}(\vec{x},\,t),\, \widehat{\mathbb{H}}] \qquad (2.203)$$

yields an operator equation corresponding to the equation (2.200). It is easy to show [42] that the action of the Hamiltonian operator $\widehat{\mathbb{H}}$ on a multi-particle state is given by

$$\langle \vec{x}_1\, \vec{x}_2\, \ldots\, \vec{x}_n|\widehat{\mathbb{H}}|\,\xi\,\rangle \;=\; [-\frac{\hbar^2}{2m}\sum_{a=1}^{n}\Delta_a \;-\; Gm^2 \sum_{a<b}\frac{1}{|\vec{x}_a - \vec{x}_b|}]\,\langle \vec{x}_1\, \vec{x}_2\, \ldots\, \vec{x}_n\,|\xi\,\rangle\,, \qquad (2.204)$$

which is consistent with the classical multi-particle Hamiltonian with gravitational pair-interactions. Evidently, the interaction Hamiltonian annihilates a single particle state $|\,\vec{x}\,\rangle := \widehat{\psi}^{\dagger}(\vec{x})\,|0\,\rangle$,

$$\widehat{\mathbb{H}}_{\mathrm{I}}\,|\,\vec{x}\,\rangle \;=\; 0\,, \qquad (2.205)$$

implying that, thanks to the appropriate normal-ordering of the operators in $\widehat{\mathbb{H}}$, the matter particles do not gravitationally self-interact. Moreover, the number operator, defined by $\hat{N} := \int d\vec{x}\ \hat{\psi}^\dagger(\vec{x})\ \hat{\psi}(\vec{x})$, commutes with the total Hamiltonian operator. In other words, the number of particles in the theory under consideration is a constant of motion: as expected, our Galilean-relativistic interaction does not lead to particle production.

2.5.4 The intuitive physical picture

It is well-known that Newtonian gravitational field does not possess any dynamical degrees of freedom of its own — they 'remain frozen' in the '$c \to \infty$' limit. On the other hand, Einsteinian gravitational radiation propagates in vacuum with the speed of light c. Considering this in the light of the local quantum field theory of Newton-Cartan gravity we have constructed here, it is rather convenient to maintain that Newtonian gravitational field also possesses propagating degrees of freedom, but it so happens that such gravitational disturbances travel instantaneously — i.e., with the Galilean-relativistic speed of light '$c = \infty$'. Indeed, it is possible to view Newton-Poisson equation $\Delta\Phi = 4\pi\,G\,\rho$ as a wave-equation for the Newtonian gravitational waves propagating with infinite speed:

$$\lim_{c\to\infty}[\Delta\Phi \ - \ \frac{1}{c^2}\frac{\partial^2\Phi}{\partial t^2} \ = \ 4\pi\,G\,\rho] \ \longrightarrow \ [\Delta\Phi \ = \ 4\pi\,G\,\rho]. \qquad (2.206)$$

Moreover, unlike the usual weak-field approach to Einstein's gravity ('$c \neq \infty$') where such *longitudinal* degrees of freedom of the gravitational field are 'gauged-away' and ignored (transverse-traceless gauge), in the theory constructed here we have been able to avoid any gauge-fixing procedure. Therefore, as far as Newton-Cartan theory is viewed as a Galilean-relativistic limit-form of Einstein's theory of gravity, the limit relation (2.206) is a better interpretation of the Newton-Poisson equation than the usual one in which one insists that there are no gravitational waves in a Newtonian theory.

Now, consider the quantized gravitational degrees of freedom. According to the above view, these also propagate with infinite speed. Moreover, if we take expressions (2.83) and (2.86) in our theory as the mass and angular-momentum operators, respectively (which are indeed the correct operators for an inertial observer), then

$$[\widehat{\mathbb{M}}, \, \widehat{\widetilde{v}}^{\alpha}] \; = \; 0 \quad \text{and} \quad [\widehat{\mathbf{J}}, \, \widehat{\widetilde{v}}^{\alpha}] \; = \; 0 \qquad (2.207)$$

imply that Newton-Cartan gravitons are the massless and spinless mediating particles between the matter particles. In other words, the theory constructed here may be viewed, in a particle interpretation, as a theory of non-self-interacting quantized Schrödinger particles producing the *longitudinal* Newton-Cartan gravitons — the massless, spin-0 exchange bosons propagating with infinite speed — and interacting through them.

2.6   Conclusion

If Einstein's gravity is viewed as a result of two physical principles, (1) strong equivalence of gravitational and inertial masses, and (2) relativization of time, then it is the second principle which prevents it from any straightforward subjugation to the otherwise well-corroborated rules of quantization. As demonstrated here, the first one by itself is completely unproblematic as far as the canonical quantization of gravity is concerned. On the other hand, the invocation of relativization of time in addition to the first principle induces the so far intractable 'problem of time' via the Hamiltonian constraint in general relativity — as is quite well-known [3]. Because of the presence of preferred foliation — whose covariant normal constitutes the kernel of the spatial metric endowing spacetime with an absolute, observer-independent notion of distant simultaneity — no such intractable constraint arises in the Hamiltonian formulation of the classical Newton-Cartan-Schrödinger theory. Consequently, we have been able to successfully and unambiguously quantize this *interacting* Galilean-relativistic field theory as an *unconstrained* Hamiltonian system in a *manifestly covariant* manner. What is more, as discussed in the Introduction, this exercise opens up a completely novel direction of research in quantum gravity: the program of the *special-relativization* of the quantum theory of Newton-Cartan gravity.

Note Added to Proof

Several years after I published this paper I learned from Roger Penrose — who, in turn, had learned from John Stachel — that in the 1930s a remarkable Soviet physicist called Bronstein had already represented the fundamental theories of physics in a map very much like the one in my Fig. 1 above. His two-dimensional map, however, neglected the Newtonian theory of gravity, which was eventually included by Zelmanov in 1967, thereby transforming Bronstein's two-dimensional picture into a full three-dimensional cube. Neither Bronstein nor Zelmanov, however, made the crucial distinction between Newton's original theory of gravity and Cartan's spacetime reformulation of it, let alone made any indication of quantization of the latter theory. In particular, Zelmanov's version of the cube (followed up by Okun in 1991 [59]) wrongly represents Newton's theory of gravity as a Galilean-relativistic limit-form of Einstein's theory of gravity. In fact, as we saw above, the true limit-form of Einstein's theory is the Newton-Cartan theory with its mutable connection field. To appreciate the importance of this crucial distinction in the contemporary context, see, for example, section 14.4.2 of Ref. [60] and references therein. For an excellent historical account of Bronstein's remarkable life and work in physics — including the Soviet history of "the cube of theories" — see Ref. [61] and references therein.

Appendix: Review of the covariant phase-space formalism

Recall that the essence of the canonical formalism for a classical system with finitely many degrees of freedom is succinctly captured by the symplectic 2-form [49, 52, 53]

$$\omega = dp_i \wedge dq^i , \qquad (2.208)$$

where q^i, $i,j = 1, ..., N$, are the generalized coordinates describing the configuration of the system, and p_j are the corresponding conjugate momenta (here Einstein's summation convention is understood). The collection of all possible values of coordinates and momenta, $(q_1, \ldots, q_N ; p_1, \ldots, p_N)$, is referred to as the *phase-space* of the system, on which the dynamical evolution is determined by a Hamiltonian function $\mathbb{H}$ via the Hamiltonian equations of motion. If we combine the p_i and q^j in a single variable Q^I, $I = 1, ..., 2N$, with $Q^i := p_i$ for $i \leq N$ and $Q^i := q^{i-N}$ for $i > N$, then we can think of ω as an antisymmetric $2N \times 2N$ matrix ω_{IJ}, with inverse ω^{IJ}, whose nonzero matrix elements are

$$\omega_{i,i+N} = -\omega_{i+N,i} = 1 . \qquad (2.209)$$

With the help of this invertible matrix ω^{IJ}, the Hamiltonian equations of motion can be succinctly expressed as

$$\frac{dQ^I}{dt} = \omega^{IJ} \frac{\partial \mathbb{H}}{\partial Q^J} , \qquad (2.210)$$

and the Poisson bracket of any two functions $U(Q^I)$ and $V(Q^I)$ as

$$\{U, V\} = \omega^{IJ} \frac{\partial U}{\partial Q^I} \frac{\partial V}{\partial Q^J} . \qquad (2.211)$$

Since the essential features of the symplectic form ω can be described in a coordinate independent manner [49, 52, 53], it can be viewed as the invariant geometric structure that underlies the definitions of Hamiltonian equations and Poisson brackets of classical mechanics. If we now view our phase-space as the cotangent-bundle T^*Z on a configuration space Z with coordinates q^i, and let

p_j be the corresponding components of a covector p at $q \in Z$, then the 2-form (2.208) is a *closed* 2-form on T^*Z,

$$d\omega = 0 \,, \tag{2.212}$$

because its components in this particular coordinate system are constant. Conversely, if ω is any closed 2-form on a $2N$ dimensional manifold $\mathcal{Z}$ such that it is non-degenerate (i.e., the matrix $\omega_{IJ}(z)$ is invertible at each point $z \in \mathcal{Z}$), then, according to Darboux's theorem [49], locally one can always introduce coordinates (p_i, q^j) on $\mathcal{Z}$, called the *canonical coordinates*, which put ω in the standard form (2.208). This immediately suggests a generalization: the cotangent-bundle form of phase-space is only a special form of phase-space (but, of course, of a fundamentally important kind) since, in general, the 'isomorphism' $\mathcal{Z} \cong T^*Z$ holds only in a local neighborhood of a point $z \in \mathcal{Z}$. In other words, globally the phase-space may not be anything like a cotangent-bundle. In general phase-space is defined as a pair $(\mathcal{Z}, \omega)$ in which $\mathcal{Z}$ is a smooth manifold, called a *symplectic manifold*, and ω, defined everywhere on $\mathcal{Z}$, is a closed, non-degenerate and real-valued 2-form called the *symplectic structure* on $\mathcal{Z}$. The 'observables' are then functions of the form $U : \mathcal{Z} \to \mathbb{R}$ generating a one-parameter family of canonical transformations on $\mathcal{Z}$. In particular, the Hamiltonian of a classical system generating the time evolution is a function $\mathbb{H} : \mathcal{Z} \to \mathbb{R}$ on the phase-space $\mathcal{Z}$ of the system, and the dynamical trajectories governed by the Hamiltonian equations of motion correspond to integral curves on $\mathcal{Z}$ of the Hamiltonian vector-field, $X_{\mathbb{H}} : \mathcal{Z} \to \mathcal{Z}$, defined by

$$X_{\mathbb{H}} \lrcorner \, \omega + d\,\mathbb{H} := 0 \,. \tag{2.213}$$

It is easy to verify that the flow of $X_{\mathbb{H}}$ preserves ω in the sense that $\mathcal{L}_{X_{\mathbb{H}}} \omega = 0$. Moreover, for a curve $Q(t)$ whose tangent vector at every point coincides with

$X_{\mathbb{H}}$, the component form of equation (2.213) in local coordinates (p_i, q^j) is precisely the equation (2.210).

The above abstract definition of phase-space quite elegantly encapsulates the important fact that it is the assignment of a symplectic structure ω on the classical phase-space $\mathcal{Z}$, rather than a particular choice of coordinates (p_i, q^j) on $\mathcal{Z}$, that is more intrinsic to the canonical formulation of a classical theory. Yet, *prima facie*, the very concept of phase-space appears to be non-covariant because phase-spaces are usually constructed by first decomposing spacetime into spacelike hypersurfaces at instants of time, and then specifying the initial data (p_i, q^j) on one of these hypersurfaces. However, thanks to the one-to-one correspondence between each dynamically allowed trajectory of a given classical system and its initial data, such a manifestly non-covariant route to phase-space is not indispensable. It is clear from equation (2.213) that each point $z \in \mathcal{Z}$ is an appropriate initial data for uniquely determining a complete Hamiltonian trajectory of the system. This establishes an isomorphism between the space of solutions to the dynamical equations and the canonical phase-space of the system, and, thereby, allows one to pull-back the symplectic structure from the phase-space to the space of solutions. As a result, we can identify our phase-space $\mathcal{Z}$ with the manifold of solutions to the equations of motion, which now we may also denote by $\mathcal{Z}$. This space of solutions $\mathcal{Z}$ equipped with the symplectic structure ω is then our desired manifestly covariant phase-space $(\mathcal{Z}, \omega)$. Of course, as allowed by Darboux's theorem, one can always choose a coordinate system on $\mathcal{Z}$, at least locally, and identify the solutions to the dynamical equations with their Cauchy data in that coordinate system; but there is no fundamental necessity to violate covariance in this manner. Besides, globally such an identification may not be possible in general, because, as noted above, globally the manifold $(\mathcal{Z}, \omega)$ may not be isomorphic to a cotangent-bundle. A cotangent-bundle can, of course, be 'polarized' [49] — i.e., foliated by the individual cotangent spaces of constant q — and, thereby, the configuration and momentum variables can be distinguished. But a general symplectic manifold may not be polarizable in this manner, and, consequently, the global identification of solutions to the equations of motion with integral curves of the Hamiltonian vector-field becomes untenable in general. Instead, the 'time-development' of a classical system is understood within this framework by interpreting the Hamiltonian flow as a mapping between *entire histories* of the system. Given a notion of time, the mapping from one dynamical trajectory to another infinitesimally distant one, whose initial data at time $t + \epsilon$ are the same as the initial data of the first trajectory at time t, is interpreted as the 'time-evolution' from t to $t + \epsilon$ generated by the Hamiltonian function $\mathbb{H} : \mathcal{Z} \to \mathbb{R}$.

The (pre)symplectic structure in the covariant description of an *infinite-dimensional* phase-space $(\mathcal{Z}, \omega)$ of a field theory is defined by a real valued function of the field variables,

$$\omega = \int_{\Sigma_t} \omega^\mu \, n_\mu \, d\Sigma_t , \qquad (2.214)$$

where Σ_t is some spacelike hypersurface, n_μ is the inverse of an 'outward-pointing' unit normal to this surface, and ω^μ — called the *presymplectic current density* — is a *conserved* and *closed* 2-form (which may be degenerate, and hence the prefix 'pre' meaning 'before the removal of degeneracy'; see below). As discussed above, if the manifold $\mathcal{Z}$ is polarizable so that we can distinguish between configuration and momentum variables, then we can express it as a cotangent-bundle of the configuration space. The above general expression for ω can then be reduced (at least in the case of a first-order action) to the standard canonical presymplectic structure,

$$\omega = \int_{\Sigma_t} dp_i \wedge dq^i \, d\Sigma_t , \qquad (2.215)$$

once a choice of global coordinates $(p_i,\, q^j)$ is made. Despite its appearance in (2.214) and (2.215), ω is independent of the choice of hypersurface Σ_t (provided that either Σ_t is chosen to be compact or the field variables on it are subjected to satisfy suitable boundary conditions at spatial infinity).

This manifest covariance of ω may require some convincing. Let us look more closely at how it can be shown by considering a dynamical theory for a collection of smooth fields $\mathcal{Q}^r(x)$ on a spacetime $\mathcal{M}$ equipped with a derivative operator ∇_μ; here r is a collective label for fields representing spacetime indices as well as internal and discrete indices. We assume that our spacetime $\mathcal{M}$ is globally hyperbolic, i.e., topologically $\mathbb{R} \times \Sigma$, where each image Σ_t of Σ for any $t \in \mathbb{R}$ is diffeomorphic to $\mathbb{R}^3$. Let Z denote the infinite-dimensional manifold constituted by the fields $\mathcal{Q}^r$ on $\mathcal{M}$. Since functions on Z are functionals of the form $f[\mathcal{Q}(x)]$, an action functional $\mathcal{I} : Z \to \mathbb{R}$ may be constructed over some measurable region of $\mathcal{M}$ such that equations of motion for a field $\mathcal{Q}^r$ can be obtained by extremizing the action under any variation $\delta\mathcal{Q}^r \equiv \mathsf{Y}^r$ of the field which vanishes on the boundary of the region. The variations $\delta\mathcal{Q}$ here are tangent vectors $\mathsf{Y} \in T_\mathcal{Q}Z$ at points in space Z corresponding to the fields $\mathcal{Q}$, the tangent spaces $T_\mathcal{Q}Z$ at $\mathcal{Q}$ are vector spaces of the variations Y, and the first variation $\delta\mathcal{I}$ of the action can be expressed as an exterior derivative of $\mathcal{I}$ in Z applied to Y:

$$\delta\mathcal{I} \;=\; d_z\,\mathcal{I}(\mathsf{Y}) \; ; \qquad\qquad (2.216)$$

i.e., the variational derivative 'δ' is equivalent to the exterior derivative 'd_z' on the space Z. Then, for a given measurable region $\mathcal{O} \subset \mathcal{M}$ with a non-null boundary $\partial\mathcal{O}$ and a collection of fields $\mathcal{Q} \in Z$, the variation of action

$$\mathcal{I}[\mathcal{Q}] \;=\; \int_\mathcal{O} dv \; \mathcal{L}(\mathcal{Q}^r,\, \nabla_\mu\mathcal{Q}^r,\, \ldots,\, \nabla_\mu\nabla_{\mu_2}\ldots\nabla_{\mu_k}\mathcal{Q}^r) \qquad (2.217)$$

is equal to

$$d_z \mathcal{I}(\mathbf{Y}) = \int_{\mathcal{O}} dv \left\{ \frac{\delta \mathcal{L}}{\delta \mathcal{Q}^r} \delta \mathcal{Q}^r + \nabla_\mu \mathcal{J}_{\mathcal{Q}}^\mu \right\}, \tag{2.218}$$

where dv is the volume element on $\mathcal{M}$ compatible with the derivative operator ∇_μ,

$$\mathcal{J}_{\mathcal{Q}}^\mu := \left\{ \frac{\delta \mathcal{L}}{\delta(\nabla_\mu \mathcal{Q}^r)} \right\} \delta \mathcal{Q}^r + \ldots + \left\{ \frac{\delta \mathcal{L}}{\delta(\nabla_{\mu\mu_2\ldots\mu_k} \mathcal{Q}^r)} \right\} \nabla_{\mu_2\ldots\mu_k} \delta \mathcal{Q}^r \tag{2.219}$$

is what is called the *presymplectic potential current density*, and the variational derivative $\frac{\delta}{\delta l}$ of a local function is defined as

$$\frac{\delta}{\delta l} := \frac{\partial}{\partial l} - \nabla_\mu \left\{ \frac{\partial}{\partial(\nabla_\mu l)} \right\} + \nabla_\mu \nabla_\nu \left\{ \frac{\partial}{\partial(\nabla_\mu \nabla_\nu l)} \right\} - \ldots \tag{2.220}$$

Under the requirement that the action $\mathcal{I}$ remains stationary for any variation $\mathbf{Y}$ of $\mathcal{Q}$ which vanishes on the boundary, the subspace $\mathcal{Z} \subset Z$ of solutions to the dynamical equations is defined by the condition $\frac{\delta \mathcal{L}}{\delta \mathcal{Q}} = 0$. Conversely, on the space $\mathcal{Z}$ consisting of fields $\mathcal{Q}$ which extremize $\mathcal{I}$, the pull-back of equation (2.218) reduces to the boundary term

$$i^* d_z \mathcal{I}(\mathbf{Y}) = \oint_{\partial \mathcal{O}} \mathcal{J}_{\mathcal{Q}}^\mu(\mathbf{Y}) \, n_\mu \, ds, \tag{2.221}$$

where $i : \mathcal{Z} \hookrightarrow Z$ is the natural embedding of the submanifold $\mathcal{Z}$ into Z, n_μ is the inverse of an 'outward-pointing' unit normal to the boundary $\partial \mathcal{O}$ as before, ds is the surface element on $\partial \mathcal{O}$, and now the tangent vector $\mathbf{Y} \in T_{\mathcal{Q}} \mathcal{Z}$ is a solution to the *linearized* equations of motion at $\mathcal{Q}$ (i.e., both $\mathcal{Q}$ and $\mathcal{Q} + \epsilon \mathbf{Y}$ satisfy the equations of motion to lowest order in ϵ). Clearly, the surface term (2.221) does not contribute to the variation of the action because $\mathcal{J}_{\mathcal{Q}}^\mu(\mathbf{Y})$ vanishes when $\mathbf{Y} \equiv \delta \mathcal{Q} = 0$ is assumed on the boundary. Nevertheless, it is precisely this term that captures the manifestly covariant essence of the canonical structure of phase-space. In order to see this, consider a volume segment $\mathcal{O}(\Sigma_{t'} - \Sigma_t) \subset \mathcal{M}$ which is bounded by two Cauchy surfaces $\Sigma_{t'}$ and Σ_t connected by a timelike world tube

$\mathcal{T}_\infty$ at spatial infinity, and define a 1-form $\theta_{\Sigma_t}(\mathsf{Y})$ on $\mathcal{Z}$ for all tangent vectors $\mathsf{Y} \in T_Q \mathcal{Z}$ by

$$\theta_{\Sigma_t}(\mathsf{Y}) := \int_{\Sigma_t} \mathcal{J}_Q^\mu(\mathsf{Y}) \, n_\mu \, d\Sigma_t \,, \tag{2.222}$$

so that in terms of this 1-form equation (2.221) becomes

$$i^* d_z \mathcal{I}(\mathsf{Y}) = \theta_{\Sigma_{t'}}(\mathsf{Y}) - \theta_{\Sigma_t}(\mathsf{Y}) + \theta_{\mathcal{T}_\infty}(\mathsf{Y}) \,. \tag{2.223}$$

Note that, in general, this *presymplectic potential* $\theta_{\Sigma_t}(\mathsf{Y})$ (as it is sometimes called in suggestive analogy with the electromagnetic vector-potential) is not unique and *depends* on the choice of hypersurface Σ_t. However, a presymplectic 2-form on $\mathcal{Z}$ can now be defined as the exterior derivative of $\theta_{\Sigma_t}(\mathsf{Y})$,

$$\omega_{\Sigma_t}(\mathsf{Y}_1, \mathsf{Y}_2) := d_z \theta_{\Sigma_t}(\mathsf{Y}_1, \mathsf{Y}_2) \,, \tag{2.224}$$

which in the light of equation 2.223 is immediately seen to behave as

$$0 \equiv i^* d_z^2 \mathcal{I} = \omega_{\Sigma_{t'}} - \omega_{\Sigma_t} + \omega_{\mathcal{T}_\infty} \tag{2.225}$$

under changes of the Cauchy surface Σ_t. Thus, in case Σ_t is non-compact, simply by a suitable choice of boundary condition (e.g., $\mathcal{J}_Q^\mu \to 0$ at spatial infinity ensuring the vanishing of the term $\omega_{\mathcal{T}_\infty}$) we obtain the desired covariant behavior of the presymplectic structure as claimed above.

In the view of equation (2.224), the closed-ness of the 2-form ω is immediate: $d_z \omega = 0$. Whereas comparisons of equations (2.214), (2.222), (2.224), and (2.225) show that the corresponding *presymplectic current density*

$$\omega^\mu(\mathsf{Y}_1, \mathsf{Y}_2) := d_z \mathcal{J}_Q^\mu(\mathsf{Y}_1, \mathsf{Y}_2) \,, \tag{2.226}$$

whose integral over Σ_t gives the 2-form ω_{Σ_t} on $\mathcal{Z}$, is conserved,

$$\nabla_\mu \omega^\mu = 0 \,, \tag{2.227}$$

thanks to the equations of motion $\frac{\delta \mathcal{L}}{\delta \mathcal{Q}} = 0$ and their linearized versions satisfied, respectively, by the fields $\mathcal{Q}$ and their variations Y. Thus, at each spacetime point x, $\omega^\mu(x)$ is a vector-valued 2-form on the space $\mathcal{Z}$ of classical solutions, but in its dependence on x, it is a conserved current density. Furthermore, in case Σ_t is non-compact, if a 4-divergence term is added to the Lagrangian, $\mathcal{L} \to \mathcal{L} + \nabla_\mu \lambda^\mu$, then the potential current density $\mathcal{J}_{\mathcal{Q}}^\mu$ changes only by an exact form $d_z \lambda^\mu$ plus an identically conserved vector density, changing the 2-form ω only by a 'surface term at infinity' [44].

It is clear from our notation in the equation (2.224) that ω can also be viewed as a skew-symmetric bilinear function $\omega : T_z \mathcal{Z} \times T_z \mathcal{Z} \to \mathbb{R}$ on the tangent vector-space $T_z \mathcal{Z}$ at some point z on the phase-space $\mathcal{Z}$. The phase-space of a linear (i.e., non-self-interacting) dynamical system is the prime example of such a symplectic vector-space with a bilinear form [54]. The field equations of linear dynamical systems are 'already linearized'; i.e., the equations of motion for such systems are linear in the linear canonical coordinates because their Hamiltonians are at most quadratic functions on $\mathcal{Z}$. Consequently, their phase-spaces, viewed as manifolds $\mathcal{Z}$ of solutions, have a natural vector-space structure, and one may identify the tangent space $T_z \mathcal{Z}$ at any point $z \in \mathcal{Z}$ with the space $\mathcal{Z}$ itself. The symplectic form $\omega(\mathcal{Q}_1, \mathcal{Q}_2)$, with $\mathcal{Q}_1, \mathcal{Q}_2 \in \mathcal{Z}$, then becomes a bilinear map on the symplectic vector-space $\mathcal{Z}$, $\omega : \mathcal{Z} \times \mathcal{Z} \to \mathbb{R}$, which is independent of the choice of point z used to make this identification.

The presymplectic 2-form ω on $\mathcal{Z}$ we have defined so far is necessarily degenerate if the action functional $\mathcal{I}$ admits any gauge-arbitrariness. Conversely, a degenerate symplectic structure would necessitate gauge-type symmetries in a theory (the issue, of course, is closely related to the question of whether or not the Cauchy problem for $\frac{\delta \mathcal{L}}{\delta \mathcal{Q}} = 0$ is well posed). However, at least in simple cases, it is possible to obtain a genuine (i.e, non-degenerate) symplectic 2-form $\tilde{\omega}$ on a reduced phase-space $\tilde{\mathcal{Z}}$ by the so-called *reduction procedure* [44, 49]. A 2-form $\tilde{\omega}$ of the pair $(\tilde{\mathcal{Z}}, \tilde{\omega})$ is said to be *non-degenerate* (or *weakly non-degenerate*, to be more precise) if $\tilde{\omega}(Y_1, Y_2) = 0$ for all $Y_2 \in \tilde{\mathcal{Z}}$ implies $Y_1 = 0$, and it is said to be *strongly non-degenerate* if the map $T_z\tilde{\mathcal{Z}} \to T_z^*\tilde{\mathcal{Z}} : Y \mapsto Y \,\lrcorner\, \omega$ is a linear isomorphism at each point $z \in \tilde{\mathcal{Z}}$ (the criteria of weak and strong non-degeneracy are clearly equivalent when $\tilde{\mathcal{Z}}$ is finite dimensional). The reduction procedure amounts to projecting down the presymplectic 2-form ω from the space $\mathcal{Z}$ of solutions of the equations of motion to the *physical* phase-space $\tilde{\mathcal{Z}}$ — the space of solutions to the equations of motion *modulo gauge-transformations*. The projection map $\mathbb{P} : \mathcal{Z} \to \tilde{\mathcal{Z}}$ which accomplishes this, assigns each element of $\mathcal{Z}$ to its gauge-equivalence class. It can be shown that every infinitesimal gauge transformation corresponds to a degenerate direction of the 2-form ω on $\mathcal{Z}$ by showing that ω is actually a unique pull-back $\mathbb{P}^*(\tilde{\omega})$ from $\tilde{\mathcal{Z}}$ to $\mathcal{Z}$ of a *non-degenerate* 2-form $\tilde{\omega}$ on the reduced phase-space $\tilde{\mathcal{Z}}$. This reduced phase-space, in turn, is just the space of orbits in $\mathcal{Z}$ of the group $\mathcal{K}$ of gauge transformations; i.e., the reduced phase-space is the quotient space $\tilde{\mathcal{Z}} := \mathcal{Z}/\mathcal{K}$, where $\mathcal{K}$ is the *characteristic distribution* of ω with fibers $\mathcal{K}_z$ at $z \in \mathcal{Z}$ defined as [49]

$$\mathcal{K}_z \;:=\; \{ Y \in T_z\mathcal{Z} \,|\, Y \,\lrcorner\, \omega = 0 \} \;\subset\; T_z\mathcal{Z} . \qquad (2.228)$$

In other words, any nonvanishing vector field $Y_\mathcal{K}$ tangent to the $\mathcal{K}$-orbits on $\mathcal{Z}$ is a degenerate direction of ω: $Y_\mathcal{K} \in V_\mathcal{K}(\mathcal{Z}) \Rightarrow \omega(Y_\mathcal{K}, Y) = 0 \; \forall \; Y \in T_z\mathcal{Z}$, where $V_\mathcal{K}(\mathcal{Z})$ denotes the set of vector fields tangent to $\mathcal{K}$. Then, for one thing, the Lie derivative of ω along the degenerate directions is always zero:

$$Y \in V_\mathcal{K}(\mathcal{Z}) \quad \Rightarrow \quad \pounds_Y \omega = d_z(Y \,\lrcorner\, \omega) + Y \,\lrcorner\, d_z\omega = 0 . \qquad (2.229)$$

Agreeably, the quotient space $\mathcal{Z}/\mathcal{K} = \tilde{\mathcal{Z}}$ is a Hausdorff manifold since the characteristic distribution $\mathcal{K}$ on $\mathcal{Z}$ is an integrable submanifold: if Y_1 and Y_2 are degeneracy vector fields with values in $\mathcal{K}$, then so is $[Y_1, Y_2]$. Clearly, the gauge-directions are eliminated in passing from $\mathcal{Z}$ to $\tilde{\mathcal{Z}}$, rendering $\tilde{\omega}$ on $\tilde{\mathcal{Z}}$ manifestly gauge-invariant. In practice, all one has to do to ensure gauge-invariance is to make sure that a given presymplectic (i.e., degenerate) structure ω on $\mathcal{Z}$ is a pull-back of the genuine (i.e., non-degenerate) symplectic structure $\tilde{\omega}$ from the reduced phase-space $\tilde{\mathcal{Z}}$: $\omega = \mathbb{P}^*(\tilde{\omega})$. This is guaranteed if and only if $\tilde{\omega}$ on $\tilde{\mathcal{Z}}$ has vanishing components in the gauge-directions:

$$Y_\kappa \lrcorner \, \tilde{\omega} \; = \; 0 \, . \tag{2.230}$$

Finally, note that, although ω is exact by definition ($\omega = d\theta$; cf. Eq. (2.224)), $\tilde{\omega}$ need not satisfy this property, because, in general, the quotient space $\tilde{\mathcal{Z}}$ could be topologically much more subtle compared to the space $\mathcal{Z}$. It is always possible, however, to find a local neighborhood on $\tilde{\mathcal{Z}}$ such that $\tilde{\omega} = d\tilde{\theta}$ within it. If $\tilde{\mathcal{Z}}$ happens to be symplectically diffeomorphic to a cotangent-bundle, then, of course, $\tilde{\omega}$ is its natural exact symplectic structure with $\tilde{\theta}$ being the standard canonical 1-form [49]. For convenience, let us illustrate the relations between various spaces we have encountered by the following diagram:

$$
\begin{array}{ccc}
Z & \xrightarrow{\;i^*\;} & \mathcal{Z} \\[1ex]
\Big\downarrow {\scriptstyle \mathbb{P}} & & \Big\downarrow {\scriptstyle \mathbb{P}} \\[1ex]
\tilde{Z} & \xrightarrow{\;i^*\;} & \tilde{\mathcal{Z}}
\end{array}
\tag{2.231}
$$

Here $\tilde{Z}$ denotes collection of all kinematically possible field configurations modulo gauge-transformations (cf. Eq. (2.173)), whereas $\tilde{\mathcal{Z}} \subseteq \tilde{Z}$ denotes collection of purely dynamically possible field configurations modulo gauge-transformations. If constraints are present, however, then $\tilde{\mathcal{Z}} \subset \tilde{Z}$, and not all kinematically possible states are dynamically possible.

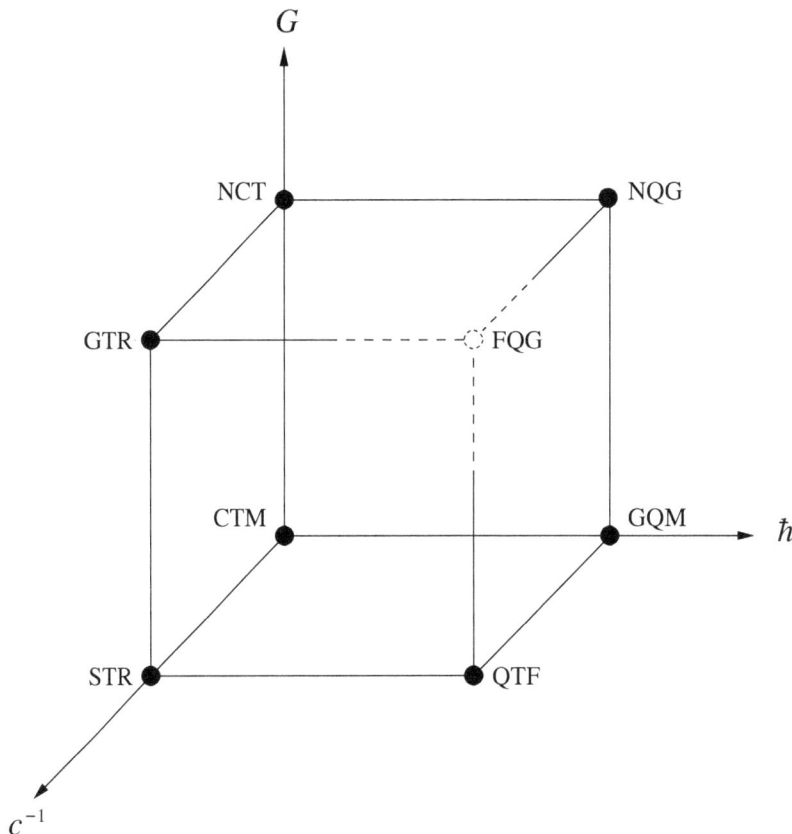

Figure 2.1: The great dimensional monolith of physics indicating the fundamental role played by the three universal constants G (the Newton's gravitational constant), $\hbar$ (the Planck's constant of quanta divided by 2π), and c (the absolute upper bound on the speed of causal influences) in various basic physical theories. These theories, appearing at the eight vertices of the cube, are: CTM = Classical Theory of Mechanics, STR = Special Theory of Relativity, GTR = General Theory of Relativity, NCT = (classical) Newton-Cartan Theory, NQG = (generally-covariant) Newtonian Quantum Gravity (constructed in this paper), GQM = Galilean-relativistic Quantum Mechanics, QTF = Quantum Theory of (relativistic) Fields, and FQG = the elusive Full-blown Quantum Gravity. If FQG turns out to require some additional fundamental constants — like the constant $\alpha' \equiv (2\pi T)^{-1}$ of the string theory controlling the string tension T, for example — then, of course, the above representation of the foundational theories would be inadequate, and additional axes corresponding to such constants α_i, $i = 1,2,...,n$, would have to be added to the diagram, making it a $3 + n$ dimensional hypercube. In that case, NQG, in particular, would be a limit-form of FQG with respect to total $n + 1$ limits, $\alpha_i \to 0$ and $c \to \infty$, in conjunction.

108

Bibliography

[1] K. Kuchař, Phys. Rev. D**22**, 1285 (1980).

[2] R. Penrose, *The Large, the Small and the Human Mind* (Cambridge University Press, Cambridge, 1997), pp 90-92.

[3] K. Kuchař, in *Proceedings of the 4th Canadian Conference on General Relativity and Relativistic Astrophysics*, edited by G. Kunstatter, D. Vincent, and J. Williams (World Scientific, Singapore, 1992), pp 211-314; and C.J. Isham, in *Integrable Systems, Quantum Groups, and Quantum Field Theories*, edited by L.A. Ibort and M.A. Rodriguez (Kluwer Academic Publishers, London, 1993), pp 157-288.

[4] É. Cartan, Ann. École Norm. Sup. **40**, 325 (1923).

[5] K. Friedrichs, Math. Annalen **98**, 566 (1927).

[6] A. Trautman, Comptes Rendus Acad. Sci. **257**, 617 (1963); and A. Trautman in *Lectures in General Relativity*, edited by S. Deser and K.W. Ford (Prentice-Hall, Englewood Cliffs, N.J., 1965), pp 1-248.

[7] P. Havas, Rev. Mod. Phys. **36**, 938 (1964).

[8] H.D. Dombrowski and K. Horneffer, Math. Zeitschr. **86**, 291 (1964).

[9] H.P. Künzle, Ann. Inst. H. Poincaré **17**, 337 (1972).

[10] W.G. Dixon, Commun. Math. Phys. **45**, 167 (1975).

[11] H.P. Künzle, Gen. Relativ. Gravit. **7**, 445 (1976).

[12] J. Ehlers, in *Grundlagenprobleme der modernen Physik*, edited by J. Nitsch et al. (Bibliographisches Institut, Mannheim, 1981), pp 65-84; J. Ehlers, in *Logic, Methodology and Philosophy of Science VII*, edited by B. Marcus et al. (Elsevier Science Publishers B.V., 1986), pp 387-403; and J. Ehlers, in *Classical Mechanics and Relativity: Relationship and Consistency*, edited by G. Ferrarese (Naples, Bibliopolis, 1991), pp 95-106.

[13] D.B. Malament, in *From Quarks to Quasars*, edited by R.G. Colodny (University of Pittsburgh Press, Pittsburgh, 1986), pp 181-201.

[14] E. Kretschmann, Annalen der Physik (Lpz.) **53**, 575 (1917).

[15] J.D. Norton, Rep. Prog. Phys. **56**, 791 (1993).

[16] J. Stachel, in *Philosophical Problems of the Internal and External Worlds: Essays on the Philosophy of Adolf Grünbaum*, edited by J. Earman, A. Janis, G. Massey and N. Rescher (University of Pittsburgh Press, Pittsburgh, PA, 1993), pp 129-160.

[17] A. Einstein, K oniglich Preussische Akademie der Wissenschaften (Berlin). Sitzungsberichte: 1030-85 (1914).

[18] P. J. E. Peebles, *The Large-Scale Structure of the Universe* (Princeton University Press, Princeton, New Jersey, 1980).

[19] U. Brauer, A. Rendall, and O. Reula, Class. Quantum Grav. **11**, 2283 (1994).

[20] C. Rüede and N. Straumann, Helv. Phys. Acta **70**, 318 (1997).

[21] J. Ehlers and T. Buchert, Gen. Relativ. Gravit. **29**, 733 (1997).

[22] E. Witten, Nuclear Physics **B** 311, 46 (1988).

[23] A. Ashtekar and M. Pierri, J. Math. Phys. **37**, 6250 (1996).

[24] J. Christian, *On Definite Events in a Generally Covariant Quantum World*, Oxford University Preprint (1994).

[25] R. Penrose, *Shadows of the Mind* (Oxford University Press, Oxford, 1994), p 339, and references therein.

[26] R. Penrose, Gen. Relativ. Gravit. **28**, 581 (1996), p 592.

[27] K. Fredenhagen and R. Haag, Comm. Math. Phys. **108**, 91 (1987).

[28] R.M. Wald, *General Relativity* (University of Chicago Press, Chicago, 1984).

[29] B. Carter and I.M. Khalatnikov, Rev. Math. Phys. **6**, 277 (1994).

[30] C. Duval and H.P. Künzle, Rep. Math. Phys. **13**, 351 (1978).

[31] D.B. Malament, Philosophy of Science **62**, 489 (1995).

[32] M. Trumper, Ann Phys. (N.Y.) **149**, 203 (1983).

[33] C. Duval, Class. Quantum Grav. **10**, 2217 (1993).

[34] C. Duval and H.P. Künzle, Gen. Relativ. Gravit. **16**, 333 (1984).

[35] J. Earman, *World enough and Spacetime* (MIT Press, Cambridge, Massachusetts, 1989), Chapter 2.

[36] E.A. Milne, Quart. J. Math. (Oxford Series) **5**, 64 (1934).

[37] S.W. Hawking and G.F.R. Ellis, *The Large Scale Structure of Space-time* (Cambridge University Press, Cambridge, 1973).

[38] R. Arnowitt, S. Deser, and C. Misner, Phys. Rev. **116**, 1322 (1959); P.A.M. Dirac, *Lectures on Quantum Mechanics* (Belfer Graduate School of Science, Yeshiva University, New York, 1964), Chapters 3 and 4; K. Kuchař, J. Math. Phys. **17**, 801 (1976); C.J. Isham and K. Kuchař, Annals of Physics **164**, 288 (1985).

[39] K. Kuchař, in *Conceptual Problems of Quantum Gravity*, edited by A. Ashtekar and J. Stachel (Birkhäuser, Boston, 1991), pp 141-168.

[40] S. De Biévre, Class. Quantum Grav. **6**, 731 (1989).

[41] S.S. Schweber, *An Introduction to Relativistic Quantum Field Theory* (Row, Peterson and Company, Evanston, Illinois, 1961), Chapter 6.

[42] L.S. Brown, *Quantum Field Theory* (Cambridge University Press, Cambridge, 1992), Chapter 2.

[43] F. Gross, *Relativistic Quantum Mechanics and Field Theory* (John Wiley, New York, 1993), Chapter 7.

[44] J. Lee and R.M. Wald, J. Math. Phys. **31**, 725 (1990).

[45] G. Barnich, M. Henneaux, and C. Schomblond, Phys. Rev. **D**44, R939 (1991).

[46] C. G. Torre, J. Math. Phys. **33**, 3802 (1992).

[47] H. F. M. Goenner, Gen. Relativ. Gravit. **16**, 513 (1984).

[48] C. Duval and H. P. Künzle, in *Semantical Aspects of Spacetime Theories*, edited by U. Majer and H.-J. Schmidt (B.I. Wissenschaftsverlag, Mannheim, 1994), pp 113-129.

[49] N. M. J. Woodhouse, *Geometric Quantization* (Clarendon Press, Oxford, 1991).

[50] C. Crnkovic and E. Witten, in *Three Hundred Years of Gravitation*, edited by S.W. Hawking and W. Israel (Cambridge University Press, Cambridge, 1987), pp 676-684.

[51] A. Ashtekar, L. Bombelli, and O. Reula, in *Mechanics, Analysis and Geometry: 200 Years after Lagrange*, edited by M. Francaviglia (Elsevier, New York, 1991), pp 417-450.

[52] R. Abraham and J.E. Marsden, *Foundations of Mechanics*, Second edition (Benjamin, Reading, Massachusetts, 1978).

[53] V. I. Arnold, *Mathematical Methods of Classical Mechanics* (Springer, New York, 1989).

[54] R.M. Wald, *Quantum Field Theory in Curved Spacetime and Black Hole Thermodynamics* (University of Chicago Press, Chicago, 1994).

[55] P.R. Chernoff, Hadronic Journal **4**, 879 (1981).

[56] R. Haag, *Local Quantum Physics* (Springer, Berlin, 1992).

[57] N. Bohr and L. Rosenfeld, Mat. Fys. Medd. K. Dan. Vidensk. Selsk. **12**, No.8 (1933).

[58] J. M. G. Fell, Trans. Am. Math. Soc. **94**, 365 (1960).

[59] L.B. Okun, Sov. Phys. Usp. **34**, 818 (1991).

[60] J. Christian, in *Physics Meets Philosophy at the Planck Scale*, edited by C. Callender and N. Huggett (Cambridge University Press, Cambridge, England, 2001).

[61] G. E. Gorelik and V. Ya. Frenkel, *Matvei Petrovich Bronstein and Soviet Theoretical Physics in the Thirties* (Birkhäuser, Basel, 1994).

Part III

Relativization of the problem :Effective Theory of Quantum Gravity

Chapter 3

Introducing Speed of Light c into Quantized N-C-S System

In this third part of the book, we introduce the speed of light into the N-C-S system and study the self-gravitating systems such as stars like neutron star, white dwarf, black hole and also the universe using a well-known many-particle Hamiltonian which is known from the early days of quantum mechanics but only recently derived [1] as representing the exactly soluble sector of quantum gravity and studied [2–4] by us from a condensed matter point of view by using a variational approach. This can also be viewed as a novel way of looking at the self-gravitating systems and it not only reproduces the results known from Einstein's General Theory of Relativity but also goes beyond by predicting certain relations and specifically the value of the cosmological constant. Instead of looking at the systems through the space-time dynamics, this theory treats the energy of the system directly. Infact from the quantum gravity point of view after quantizing the GTR in the $c \to \infty$ (Newton-Cartan theory with spatial vanishing curvature), we have explicitly the above known Hamiltonian. We then special relativize the problem by using Mach's principle in case of the universe and through the Schwarzschild radius in case of other self-gravitating systems such as neutron star, white dwarf etc. Figure 1.1 represents the underling physical theories relating to quantum theory of gravity (denoted by FQG) which is still an elusive one. But ours is a realization (denoted by a circle with a cross inside) of this FQG in the vanishing spatial curvature limit with special-relativization done on the problem which is described by the quantum many-particle Hamiltonian obtained by quantizing the Newton-Cartan theory which we write as NQG (see Fig. 1.1). We give an abstraction of Fig. 1.1 in Fig. 3.1. The special relativization is carried out by invoking the relation

$$\frac{GM}{Rc^2} \approx 1. \tag{3.1}$$

In case of universe, R is the radius and the relation represents what is known as Mach's principle and in case of stars, R is the schwarzschild radius of the corresponding stars.

3.1 Self-gravitating systems

An expanding system of self-gravitating particles can be described by the Hamiltonian which is given as

$$H = -\sum_{i=1}^{N} \frac{\hbar^2}{2m} \nabla_i^2 + \frac{1}{2} \sum_{i=1,\, i \neq j, j=1}^{N} \sum^{N} v(|\,\vec{X}_i - \vec{X}_j\,|) - \sum_{i=1}^{N} \Lambda mc^2 |\vec{X}_i|^2 \tag{3.2}$$

where $v(|\,\vec{X}_i - \vec{X}_j\,|) = -g^2/|\,\vec{X}_i - \vec{X}_j\,|$, with $g^2 = Gm^2$, G being the universal gravitational constant and m the mass of the constituent particles. In case of our universe, the last term in the above Hamiltonian describes the expansion due to the Dark Energy, which can be taken as the nonrelativistic limit of the Λ term known as Einstein's cosmological constant term in General Theory of Relativity. In this section we look into some of the properties of finite many-body systems such as neutron stars, white dwarfs etc. where the gravitational attraction is the

117

dominant interaction. As a generalization of our method of approach used for the study of stars, we have made an attempt to deal with the evolution of the present universe also. Since the stars are mostly known to be constituted of protons, neutrons and electrons, they are to be considered as systems of fermions. The reason for choosing this class of Fermi systems is that they are not so well understood and hardly there has been any attempt to study them from a condensed matter point of view applying quantum mechanics. As far as the evolution of the present universe is concerned , we visualize it to be constituted of some fictitious particles each of mass m that are of fermionic in nature. calculations of binding energies of systems of self-gravitating particles like stars after they finish up their nuclear fuel in their cores is extremely difficult. It has been shown by Fisher and Ruelle [5] that in order to establish a rigorous basis for the statistical mechanics of an infinite system of interacting particles it is necessary that the relevant forces acting on the system must be of a saturating character. For this case and for a finite system, the total energy ought to posses a lower bound and it is to be an extensive quantity, that is, proportional to the number of particles in the system. If, on the other hand the forces are not saturating character, the binding energy per particle increases indefinitely with the particle number and it becomes obviously impossible to define the usual thermodynamic variables for an infinite system. However, in case of systems governed by Coulomb forces it has been proved by Dyson and Lenard [6] through a splendid analysis that there is saturation if the particles having charges of a given sign belong to a finite number of fermion species. This is how one could explain the stability of ordinary matter. Even with the above analysis, it has not been possible yet to understand the stability of matter by including the relativistic features of electromagnetic interaction. The case of universal gravitational interaction is also difficult to study from a statistical mechanics point of view although it is very interesting since this is relevant for the matter in bulk on an astronomical scale. The difficulty with the gravitational forces is due to the fact that they are of long range nature and purely attractive. Hence they do not saturate. Nonextensive nature of thermodynamic functions of classical systems with gravitational interactions in one and two dimensions was investigated by Salzberg [7] sometime back. In a later development, rigorous inequalities were derived by Levy-Leblond [8] for the ground state energy of nonrelativistic quantum mechanical system of N particles governed by gravitational interaction. In this work it was shown by him that the binding energy per particle increases like N^2 for Bose system, whereas for Fermi system it varies as $N^{4/3}$. He was able to obtain both an upper and a lower bound for the binding energy of the system which, for large N, were given as

$$-(0.5)N^{7/3}(\frac{mg^4}{\hbar^2}) \leq E_0 \leq -(0.001055)N^{7/3}(\frac{mg^4}{\hbar^2}) \qquad (3.3)$$

We will see later that our calculated energy of the system does satisfy this limit. As an extension of this theory to system consisting of N negative light fermions each of mass m and N positive heavy particles with mass M interacting via Newtonian and Coulomb forces, Levy-Leblond has shown that the total energy is no longer bounded from below for the number of particles in the system N exceeding a critical value. This is done by accounting for relativistic effects in the kinetic energy term. Such a theory when applied to white dwarfs reproduces the Chandrasekhar limit [9] beyond which gravitational collapse occurs. Unfortunately this theory cannot be generalized to a system like neutron star where the particles are heavy and a form a degenerate Fermi gas, or to situations where relativistic treatment of the gravitational interaction itself becomes necessary. In a later work, Ruffini and Bonazzola used a self-consistent field method [10] to study the equilibrium configuration of a system of self garvitating particles (scalar boson or spin half fermions) in the ground state without going for the traditional equation of state or perfect fluid approximation. In order to apply this theory to neutron stars they had to introduce into their theory corrections due to special and general theory of relativity. With these they could derive a critical number for the particles (bosons or fermions) above which mass, gravitational collapse occurs. We study the above self-gravitating systems such as neutron star, white dwarf, black hole and also the universe quantum mechanically [2–4] in this book

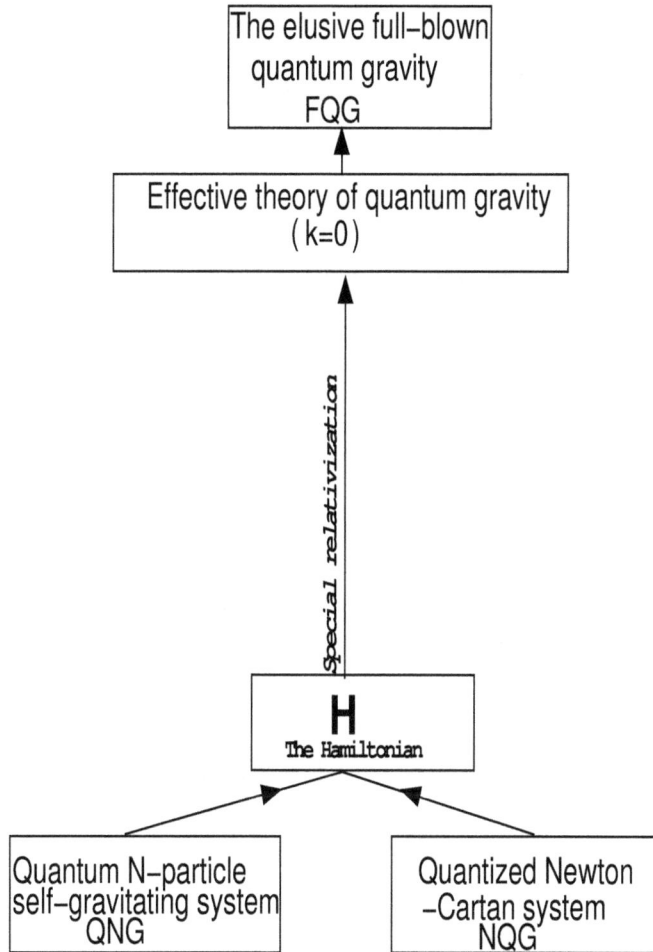

Figure 3.1: This is an abstraction of the Fig. 1 of chapter 1. The Hamiltonian which one gets after quantizing Newtonian-Cartan-Schrödinger system denoted as NQG or writing the quantum mechanical Hamiltonian for the Quantum N-Particle self-gravitating system denoted as QNG is the Hamiltonian H which when relativized with curvature $k = 0$ gives the Effective Theory of Quantum Gravity ETQGK0. ETQGK0 is a subset of the theory of elusive full blown quantum gravity.

Bibliography

[1] J. Christian, Phys. Rev. D **56**, 4844 (1997).

[2] S. Mishra, Int. J. Theor. Phys. **47**, 2655 (2008).

[3] D. N. Tripathy and S. Mishra, Int. J. Mod. Phys. D **7**, 6, 917 (1998).

[4] D. N. Tripathy and S. Mishra, Int. J. Mod. Phys. D **7**, 3 , 431 (1998).

[5] M. E. Fisher and D. Ruelle , J. Math. Phys. **7**, 260 (1966).

[6] F. J. Dyson and A. Linard, J. Math. Phys. **7**, 423 (1967).

[7] A. M. Salzberg, J. Math. Phys. 6, 158 (1965).

[8] J. M. Levy-Leblond, J. Math. Phys. **10**, 806, (1969).

[9] S. Chandrasekhar, *Mon. Not. Roy. Astron. Soc.* **91**, 456 (1931).

[10] R. Ruffini and S. Bonazzola, Phys. Rev. **187** , 1767 (1969).

[11] E. Cartan, Ann. Ecole Norm. Sup. **40**, 325 (1923).

[12] G. Contopoulos and D. Kotsakis, *Cosmology* (Springer, Heidelberg, 1987).

Chapter 4

Stars and Gravitational Collapse

In this chapter, we study compact objects like stars, neutron stars and white dwarfs by formulating the problem as an application of the quantum-many particle Hamiltonian discussed in last chapter. We, have succeeded in obtaining a nonrelativistic quantum mechanical derivation for the ground state binding energy of a system of self-gravitating particles by making a suitable choice for the single particle density. This has enabled us to arrive at a compact expression for the radius of astronomical objects like stars. Using the present theory, we have been able to estimate the critical mass of a neutron star beyond which black hole formation takes place. The present derivation of the Chandrasekhar limit for the white dwarf formation is based on introducing a radius equivalent to the Schwarzschild radius at the region of interface between the white dwarfs and neutron stars. From all these successes we feel that our choice for the single particle density of a system of self-gravitating particles is correct.

4.1   Introduction

It was first pointed out by Chandrasekhar [1] and then,independently, by Landau [2] ,long back that a degenerate system composed of a large number of self gravitating particles will necessarily undergo gravitational collapse if the particle number exceeds certain critical value. This happens after the stars finish up their nuclear fuel. Soon after this, Chandrasekhar [3] made the momentous discovery regarding the life history of certain stars, according to which the stars with masses M less than $\approx 1.4M_\odot$, $M_\odot$ being the solar mass, evolve in the same way as the sun after the nuclear power in their cores gets exhausted. When this happens, they contract to white dwarfs. In a white dwarf star, the assembly of the free electrons within the star, which usually forms a degenerate Fermi gas exerts sufficient outward pressure to counteract the inward gravitational pull. A star like our sun is said to lie on the main sequence of the Hertzsprung-Russel (HR) diagram [4], since it has still the source at its core for the generation of energy. In the distant future, it is also supposed to evolve to become a red giant and then finally to a white dwarf. Coming to the case of stars having masses less than about three times the solar mass, they may condense even more as they collapse such that their density becomes comparable to that of the nucleons inside the atomic nuclei. At this stage, the electrons and protons react by inverse β-decay and form neutrons. This is how the neutron stars are formed. These are the compact objects having a dominance of neutrons in their interiors. As such, in them, the outward pressure arises from the degenerate neutrons. Lastly, one comes across the most interesting case of stars that are having masses more than three times the solar mass. In such cases, the collapse is complete and they lead to the formation of the so called black holes. As the name implies, the black holes trap light and material particles falling on them and also prevent these from getting out of them. This is due to the fact that gravity is very strong inside the black holes. Mathematically, when the radius of a neutron star becomes less than a certain limit called the Schwarzschild radius $R_s = (\frac{2GM}{c^2})$, M being the mass of the star, G Universal Gravitational Constant and c, the velocity of light, then only it can become a black hole. Since, it is the gravitational attraction among the particles in a gravitating system which makes it to collapse, this amounts to an enormous increase in density and the temperature at its central region. For a star becoming a black hole ($M \geq 3M_\odot$), the whole star enters the horizon and ends up as a singularity at the centre. That is, the centre of a black hole is considered to be a mathematical singularity where the matter is supposed to have infinite density (Fig. 4.1).

We, in the present work, have tried to calculate the binding energy of a self-gravitating system of particles by treating them as fermions. As far as the evaluation of the total kinetic energy of the system is concerned, it is done within the Thomas-Fermi approximation (TF) [5]. The potential energy of the system is being evaluated within the so called Hartree approximation. The form of the single-particle density for the system used by us in the present calculation is such that it has a singularity at the origin. Unlike the earlier calculations, the method used here is a nonrelativistic quantum mechanical derivation based on Newtonian mechanics. The most interesting result of the present theory is that it gives rise to a compact expression for the radius of a star, following which we are able to obtain a limiting value for the critical mass of a neutron star in a natural way beyond which the black hole formation takes place. A further generalization of the present work to the case of white dwarfs enables us to derive the so called Chandrasekhar limit. In sec.2 of this chapter we have presented the mathematical formulation of our theory. Sec.3 is devoted to the various situations which lead to the formation of neutron stars, and black holes and the derivation of the well-known Chandrasekhar limit for the white dwarfs. In sec.4 a brief discussion of the results of the present theory is given.

4.2 Hamiltonian of stars

The Hamiltonian for the system of N gravitating particles each of mass m interacting through a sum of pair-wise gravitational interactions is written as

$$H = \sum_{i=1}^{N} -\frac{\hbar^2 \nabla_i^2}{2m} + \frac{1}{2} \sum_{i=1}^{N} \sum_{j=1}^{N} v(| \vec{X}_i - \vec{X}_j |) \tag{4.1}$$

where $v(| \vec{X}_i - \vec{X}_j |) = -g^2 / | \vec{X}_i - \vec{X}_j |$, with $g^2 = Gm^2$, G being Newton's Universal gravitational constant. Using this, the ground state energy of the system at zero temperature is given as [5]

$$E_0 = < H > = < KE > + < PE > \tag{4.2}$$

In the present case we confine ourselves to the system of neutrons only. Since the wavefunction of a neutron star is not known, we proceed [6] to evaluate the total kinetic energy of the system using a model density distribution function for particles (neutrons) within the neutron star. The particles being fermions, we use the Thomas-Fermi formula for calculating the total kinetic energy of the system. For an infinite many-fermion system, the average particle density and the Fermi momentum of a particle are related to each other as:

$$n = \frac{k_F^3}{3\pi^2}. \tag{4.3}$$

For a finite system like the star, since the density distribution is a function of radius vector $\vec{r}$, the fermi energy of a particle is supposed to be dependent on r. For an infinite many-fermion system, the total kinetic energy of the system [7] is given as

$$< KE >_{inf} = \frac{3}{5} n \epsilon_F = \frac{3}{5} n (3\pi^2 n)^{2/3} \frac{\hbar^2}{2m}$$
$$= \frac{3\hbar^2}{10m} (3\pi^2)^{2/3} n^{5/3}, \tag{4.4}$$

126

In analogy with Eq.(4.4), for a finite system [7], we now evaluate the kinetic energy. Assuming the particles to be fermions,the total kinetic energy of the system $< KE >$ has been evaluated within the Thomas-Fermi approximation,whose expression is given as

$$< KE >= (\frac{3\hbar^2}{10m})(3\pi^2)^{2/3} \int d\vec{X} [\rho(\vec{X})]^{5/3} \tag{4.5}$$

and the total potential energy $< PE >$ is written as

$$< PE >= -(\frac{g^2}{2}) \int d\vec{X} d\vec{X}' \frac{1}{|\vec{X} - \vec{X}'|} \rho(\vec{X}) \rho(\vec{X}') \tag{4.6}$$

In order to evaluate the above integrals, we assume that all the particles within the system (which are identical in nature) are described by some kind of distribution. The trial single-particle density $\rho(\vec{X})$ we choose is of the form:

$$\rho(\vec{X}) = A[exp(-x^\alpha)]/x^{3\alpha}, \tag{4.7}$$

where $x = (r/\lambda), r = | \vec{X} |$ and A is the normalization constant, such that

$$\int \rho(\vec{X}) d\vec{X} = N \tag{4.8}$$

The index $'\alpha'$ has been adjusted in order to bring the expression for the binding energy to have the correct dependence with the particle number. Besides, the convergence of the integrals is also to be satisfied. As one can notice from above, $\rho(\vec{X})$ is singular at $r = 0$. Having thought of a singular form of single-particle density, we have tried with a number of singular form of single particle densities of the kind $\rho(\vec{r}) = A\frac{exp[-(\frac{r}{\lambda})^\nu]}{(\frac{r}{\lambda})^{3\nu}}$ where $\nu = 1, 2, 3, 4...or$ $\frac{1}{2}, \frac{1}{3}, \frac{1}{4}, ...$ Integer values of ν are not permissible because they make the normalization constant infinite. Out of the fractional values, $\nu = \frac{1}{2}$ is found to be most appropriate, because, it will be shown here that it gives the expected upper limit for the critical mass of a neutron star [6] beyond which black hole formation takes place. Also because if ν goes to zero (like $1/n$, $n \to \infty$), $\rho(r)$ would tend to the case of a constant density as found in an infinite many-fermion system. In view of the arguments put forth above, one will have to think that the very choice of our $\rho(r)$ is a kind of ansatz in our theory, which is equivalent to the choice of a trial wave function used in the quantum mechanical calculation for the binding energy of a physical system following variational techniques. In case of a black hole it simply means that the centre of a black hole is a mathematical singularity where matter has infinite density. As far as the universe is concerned, a singularity in the particle density at origin is thought to be related to the so called Big Bang theory, which is being assumed to be the most important current theory for the origin of the Universe. There have been a few most important advances in this direction by Hawking and Penrose [4] who have shown that any model of the Universe which has the observed characterstics of approximate homogeneity and isotropy must start from a singularity. Even, Einstein's General Theory of Relativity (GTR) which when applied to cosmology accounts for such an initial singularity of the Universe.

Evaluations of the integrals in Eq.(4.5-4.6) have been made taking $\alpha = \frac{1}{2}$. With help of the above choice for $\rho(X)$, the expression for the total energy, $E_0(\lambda)$ of the system of N self-gravitating particles is obtained as

$$E(\lambda) = \left(\frac{12}{25\pi}\right)\left(\frac{\hbar^2}{m}\right)\left(\frac{3\pi N}{16}\right)^{5/3}\frac{1}{\lambda^2} - \left(\frac{g^2 N^2}{16}\right)\frac{1}{\lambda} \qquad (4.9)$$

Minimizing this with respect to λ, it is found that the minimum occurs at

$$\lambda = \lambda_0 \simeq \left(\frac{\hbar^2}{mg^2}\right) \times (2.023764)/N^{1/3} \qquad (4.10)$$

Evaluating Eq.(4.9) at $\lambda = \lambda_0$, the total binding energy of the system is found as

$$E_0 \simeq -(0.015442)N^{7/3}\left(\frac{mg^4}{\hbar^2}\right) \qquad (4.11)$$

128

Considering the case of the two-particle system (N=2), from Eq.(4.11), we find

$$E_0 = -(0.077823)\left(\frac{mg^4}{\hbar^2}\right) \tag{4.12}$$

This is seen to be quite high compared to the actual binding energy of the two-body system whose value is (-0.25) $\left(\frac{mg^4}{\hbar^2}\right)$. Comparing the two results, one should not consider Eq.(4.11) to be a drawback of the present theory,because it is supposed to be very accurate for very large N. Looking at Eq.(4.11), we find that E_0 varies as $N^{7/3}$ where N is the particle number. Such a dependence of the binding energy for the system on N was also found by Levy-Leblond [8] by assuming the particles to be fermions. By this, he was able to obtain both an upper and a lower bound for the binding energy of the system which, for large N, were given as

$$-(0.5)N^{7/3}\left(\frac{mg^4}{\hbar^2}\right) \le E_0 \le -(0.001055)N^{7/3}\left(\frac{mg^4}{\hbar^2}\right) \tag{4.13}$$

Anyway, comparing our result, as shown in Eq.(4.11), with Eq.(4.13), we find that it does not violate the inequalities established by Levy-Leblond [8].

4.3 Formation of compact objects

4.3.1 Neutron stars and black holes

Before we proceed to make an estimate of the critical mass of a neutron star beyond which black hole formation should take place,we have to first know about the radius of a star. It must be noted that the size of any compact object (either an atom or a star) is not well defined in quantum theory. The justification regarding the identification of the radius R_0 of a star with $2\lambda_0$ follows from the consideration of the so called tunneling effects used in quantum mechanics. Classically,it is known that a particle has a turning point where the potential energy becomes equal to the total energy [9]. Since the kinetic energy and therefore the velocity are equal to zero at such a point,the classical particle is expected to be turned around or reflected by the potential barrier. For example,considering the case of an electron in the hydrogen atom ground state such classical turning point occurs where the potential $V(r) = -e^2/r = E_{total} = -e^2/2a_0$; that is at $r = 2a_0$. Quantum mechanically,the probability distribution $r^2\rho(r)$ has a non-zero value for $r > 2a_0$; that is, the electron has access to the region $r > 2a_0$ which is forbidden by classical theory. Such penetration or tunneling into or through the potential energy barriers is typical of quantum theory results. If the electron had a value of $r > 2a_0$, then its kinetic energy would have to be negative to satisfy the condition $E_{total} = T + V$,with $V > E_{total}$. Since negative kinetic energy is physically absurd, $r = 2a_0$ is to be identified as the classical radius. Using the above idea, from the present theory one can easily see that at $\lambda = 2\lambda_0$, the potential energy of the system becomes equal to the the total energy, there by proving that the radius of the star $R_0 = 2\lambda_0$.

In order that a neutron star, after it finishes up all its nuclear fuel at the centre, would form a black hole, one must have

$$R_0 \leq R_s, \tag{4.14}$$

where

$$R_s = \left(\frac{2GM}{c^2}\right) \tag{4.15}$$

is the Schwarzschild radius [4] of the corresponding black hole. Following Eq.(4.10) and Eq.(4.14), one finds that the number of nucleons N in the star satisfies the inequality

$$N \geq 1.696758N_1, \tag{4.16}$$

where

$$N_1 = (\frac{\hbar c}{Gm_n^2})^{3/2}, \tag{4.17}$$

m_n being the mass of a neutron. The equality sign in Eq.(4.16) refers to the critical value, denoted by N_{C_1}, for the number of particles in a neutron star beyond which black hole formation takes place. A numerical estimation of N_{C_1} gives

$$N_{C_1} \simeq 1.70 N_1 \approx 3.73 \times 10^{57} \tag{4.18}$$

From Eq.(4.17) it follows that

$$N_1^{2/3} = (\frac{\hbar c}{Gm_n^2}) = \left(\frac{Planck\ mass}{nucleon\ mass}\right)^2 \tag{4.19}$$

Looking at the result given in Eq.(4.18), one finds that this is in very good agreement with the well known result for the number of nucleons in a typical star, as estimated earlier [10]. Using this, one also finds that

$$M_{C_1} = N_{C_1} m_n \simeq 3.122134 M_\odot, \tag{4.20}$$

where $M_\odot = 2 \times 10^{33} g$, is the mass of the sun. Thus, we find that for neutron stars more massive than $\approx 3 M_\odot$, the collapse is complete and these are the stars which lead to the black holes [4]. Now, corresponding to $N = N_{C_1}$, we calculate the radius of a neutron star, which gives

$$R_0 = 2\lambda_0 \leq 3.39352(\frac{\hbar}{m_n c})(\frac{\hbar c}{Gm_n^2})^{1/2} \tag{4.21}$$

This is the same result as found earlier by Shapiro and Teukolsky [11] (ST). A numerical estimate of Eq.(4.21) gives $9.25 \times 10^5 cm$ compared to the value of $3 \times 10^5 cm$ quoted by ST.

4.3.2 White dwarfs and the Chandrasekhar limit

In view of the result shown in Eq.(4.20), it is apparent that if the mass of a star is less than $\approx 3M_\odot$, but not too low, it must remain as a neutron star. At this stage, one is likely to ask, is there any lower bound on the mass of a neutron star ? In order to answer this question, we imagine of a radius, denoted by R'_s, equivalent to the Schwarzschild radius, upto which the neutron star is likely to exist. Above this R'_s, one no longer talks of a neutron star. Rather, one has to speak of a white dwarf, provided the mass of the star is less than the Chandrasekhar limit at the time when its nuclear fuel gets exhausted. Mathematically, we write down the expression for the R'_s as

$$R'_s = \frac{2GM}{<\vec{v}^2>} \tag{4.22}$$

As one can see from above, R'_s has been written in a fashion similar to the Schwarzschild radius except for the fact that the c^2 factor in the Schwarzschild radius has been replaced by the average of the velocity square $<\vec{v}^2>$. The quantity $<\vec{v}^2>$ is to be here understood as square of the escape velocity of a particle from a neutron star. Quantitatively, we choose $<\vec{v}^2>$ as

$$<\vec{v}^2> = c^2\left(\frac{m_e}{m_n}\right)^\eta \tag{4.23}$$

where the value of the exponent η in the above equation is to be adjusted in order to reproduce the value 2 for the degree of ionization for heavy nuclei. In doing this, the so called Chandrasekhar limit [3] for the mass of a white dwarf ($M \approx 1.44M_\odot$) is obtained in a natural way. In order to show this, we now consider the following inequality,

$$R_s < R_0 < R'_s = \left[\frac{2GM}{c^2}\right]\left(\frac{m_n}{m_e}\right)^\eta, \tag{4.24}$$

Analyzing Eq.(4.24), for the case $R_0 < R'_s$, we obtain

$$N = N_{C_2} \geq 1.696757\left(\frac{\hbar c}{Gm_n^2}\right)^{3/2}\left[\left(\frac{m_e}{m_n}\right)^{3/4}\right]^\eta \tag{4.25}$$

132

Following this, we write

$$M_{C_2} = m_n N_{C_2} \geq 3.126 M_\odot \left[3.5613 \times 10^{-3} \right]^\eta \qquad (4.26)$$

Using the above equation, we now go on varying η. For each value of η, we try to calculate the degree of ionization μ_e using the relation [12]

$$\mu_e^2 = 5.83 \left(\frac{M_\odot}{M_{C_2}} \right) \qquad (4.27)$$

It can be easily seen that only when $\eta = 0.137271$, μ_e becomes 2.01. For heavy nuclei, it has been known that μ_e, which is being interpreted as the degree of ionization has a value close to 2. Now, corresponding to the above η, we find that

$$M_{C_2} \simeq 1.44 M_\odot, \qquad (4.28)$$

the well known Chandrasekhar limit [3]. A further justification regarding our above choice of R'_s is given in sec.4. Thus, the mass of a neutron star happens to be such that $M_{Ch} \leq M^{NS} \leq 3.12 M_\odot$. For a star having masses $M < M_{Ch}$, the formation of white dwarfs should take place after such a star finishes up all its nuclear energy.

In order to calculate the radius of a white dwarf star, one has to consider the fact that in these stars, the outward pressure is due to the degenerate electrons rather than due to the neutrons as is the case with the neutron stars. Therefore, in white dwarfs, it is this outward electron pressure which is counterbalanced by the inward gravitational pull arising out of the protons. While generalizing the present calculation to white dwarfs, we ignore the effect of the gravitational forces between the electrons and electrons and between electrons and protons, as these are negligibly small. This is justified considering the fact that the proton mass is very high compared to the electron mass. Thus the mass 'm' that appears in the kinetic energy term in Eq.(4.1) should now represent the electron mass

m_e and the symbol g^2 that appears in the interparticle potential term should, as before, be given as $g^2 = Gm_n^2$, m_n being the mass of a neutron. With these modifications, the expression for R_0^{WD} is obtained as

$$R_0^{WD} \simeq \left(\frac{\hbar^2}{Gm_em_n^2}\right)4.047528/N^{1/3} \tag{4.29}$$

R_0^{WD} should be such that its value has to be greater than R_s'. It can be easily verified that for masses $M \leq 1.44 M_\odot$, $R_0^{WD} > R_s'$. For $M = 1.0 M_\odot$, we have calculated the radius of the white dwarf using Eq.(4.29). This gives $R_0^{WD} \approx 2.49 \times 10^9$ cm. which is in close agreement with the value estimated by others [11]. For this mass, $R_s' \approx 0.832 \times 10^6 cm$; thus showing that $R_0^{WD} > R_s'$. Using the above value of R_0^{WD}, we have estimated the mass density inside a white dwarf of mass $M = 1.0 M_\odot$. This gives $\rho^{WD} \approx 3.1 \times 10^4 g/cm^3$, which is again of the right order of magnitude as reported by others [13]. Using the above value of ρ^{WD}, the density of particles within a white dwarf star is found to be $\approx 1.80 \times 10^{28}\ cm^{-3}$. It is because of such a high value for the particle density, the effects of the Pauli exclusion principle becomes important in such stars and hence, the matter in such a state is considered to be quantum mechanically degenerate.

Since R_0^{WD} is also supposed to be larger than R_s, this gives rise to the fact that

$$N \geq 1.696757\left(\frac{\hbar c}{Gm_n^2}\right)^{3/2}\left(\frac{m_n}{m_e}\right)^{3/4} \tag{4.30}$$

Following this, one obtains

$$R_0 \geq 0.5184(\frac{\hbar}{m_ec})(\frac{\hbar c}{Gm_n^2})^{1/2} \tag{4.31}$$

This is the well known relation as obtained before [11]. The expression in the right hand side of the above equation when evaluated gives $\approx 2.6 \times 10^8 cm$. For $M = 1.0 M_\odot$, the estimated value of R_0^{WD} actually satisfies the above inequality. Now, consider the case of a neutron star of mass $M = 1.5 M_\odot$. Following Eq.(4.10), its radius becomes $R_0 = 1.18 \times 10^6$ cm. Using this, the matter density inside such a star is found to be $\rho^{NS} \approx 4.3 \times 10^{14} g/cm^3$. This being of the same order as the mass density within an atomic nucleus, one is justified to call them as neutron stars. In the black hole state ($M \simeq 3.2 M_\odot$), the radius of the corresponding neutron star becomes $R_0 = 9.18 \times 10^5 cm$. Its Schwarzschild radius R_s is found to be $3(\frac{M}{M_\odot}) km \simeq 9.6 \times 10^5 cm$. Thus, one finds that for a black hole, $R_0 < R_s$. This is what is expected to happen for neutron stars having masses $M \geq 3.12 M_\odot$.

Coming back to the case of a star in the white dwarf stage with a mass $M \simeq 1.0 M_\odot$, we have estimated the mean temperature throughout the body of such a star by requiring that the thermal kinetic energy of the star be equal to its gravitational potential energy. Using the present theory, we have calculated the total binding energy of a white dwarf star of mass $M = 1 M_\odot$, using the expression

$$| E_0 | = 0.015442 N^{7/3} \frac{G^2 m_e m_n^4}{\hbar^2}, \qquad (4.32)$$

This gives

$$| E_0 | \approx 0.67 \times 10^{49} \; ergs \qquad (4.33)$$

Comparing our result with those estimated earlier [14], we find that the agreement is extremely good. Leaving aside the details of the composition, as far as the star like the sun is concerned, which is at present a star on the main sequence, it can be considered to resemble with a white dwarf after its nuclear fuel gets exhausted. Using virial theorem, which tells that the sum of the potential energy and twice the kinetic energy of a self-gravitating system is zero [4], we obtain

$$| E_{pot} | \simeq 1.34 \times 10^{49} ergs, \qquad (4.34)$$

Following Eq.(4.34), we now calculate the value of the potential energy per gramme, and then equate it with the mean thermal kinetic energy, $\frac{1}{2} v^2$, per gramme of a particle (Hydrogen atom) inside the white dwarf. This gives the mean thermal velocity v of the particle as $v \simeq 1.6 \times 10^3 km/sec$, and hence, the corresponding mean temperature becomes $\sim 5.5 \times 10^7$K . The central temperature of a white dwarf has to be much more than the above value. As far as the sun is concerned, since its binding is found to be less than that of a white dwarf of the same mass [14], it is expected that the mean temperature of the sun has to be less than that of a white dwarf. The same is true for the central temperature which, for the sun, has a value $\sim 2 \times 10^7 K$.

4.4   Discussion

As seen before, in order that a star can go to white dwarf state after its nuclear fuel at the core gets exhausted, it must have a mass less than $\approx 1.44 M_{\odot}$, the well known Chandrasekhar limit. To arrive at this result, we have introduced a radius R'_s, equivalent to the Schwarzschild radius R_s, such that $R_0 < R'_s$, where we have defined $R'_s = \frac{2GM}{<\vec{v}^2>}$, having $< \vec{v}^2 > = c^2(\frac{m_e}{m_n})^\eta$, which is being interpreted as the escape velocity of a particle from the surface of the neutron star. In order that the above inequality is to be satisfied, one must have $M < M_{Ch} \approx (1.44) M_{\odot}$, which corresponds to an $\eta = 0.137271$. For such an η, we reproduce a value for the escape velocity $v(v \approx 0.62c)$ of a particle from the surface of the neutron star which is found to be of the right order of magnitude [15]. This also gives the correct result for the degree of ionization, $(\mu_e \approx 2)$.

4.5   Summary

In this chapter, we study compact objects like stars, neutron stars, white dwarfs and black holes by formulating the problem as an application of the quantum-many particle Hamiltonian discussed in last chapter. By making an intuitive choice for the single-particle density of a system of N self-gravitating particles, without any source for the radiation of energy, we have been able to calculate the binding energy of the system by treating these particles as fermions. Our expression for the ground state energy of the system shows a dependence of $N^{7/3}$ on the particle number, which is in agreement with the results obtained by other workers. We also arrive at a compact expression for the radius of a star following which we correctly reproduce the nucleon number to be found in a typical star. Using this value, we obtain the well-known result for the limiting value of the mass, M, of a neutron star ($M \simeq 3.12 M_\odot$, $M_\odot$ being the solar mass) beyond which the black hole formation should take place. Generalizing the present calculation to the case of white dwarfs, we have been able to obtain the so called Chandrasekhar limit for the mass, M_{Ch}, ($M_{Ch} \simeq 1.44 M_\odot$) below which the stars are expected to go over to the white dwarf state. We reproduce this by introducing a radius, equivalent to Schwarzschild radius, at the interface of the neutron stars and white dwarfs. This is justified by considering the fact that it gives rise to the correct value for the degree of ionization μ_e ($\mu_e \approx 2$) for heavy nuclei.

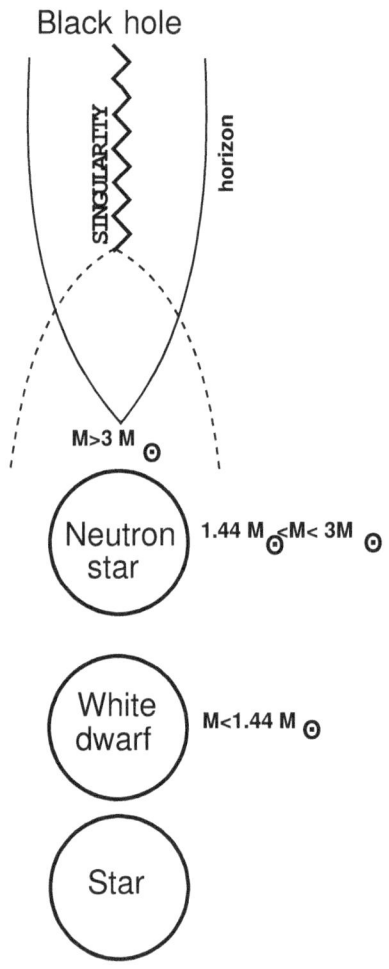

Figure 4.1: A schematic figure showing the stars with different masses leading to a black hole

Bibliography

[1] S. Chandrasekhar, *Mon. Not. Roy. Astron. Soc.* **91**, 456 (1931).

[2] L. Landau, Phys. Z. USSR **1**, 285 (1932).

[3] S. Chandrasekhar, *Mon. Not. Roy. Astron. Soc.* **95**, 207 (1935).

[4] G. Contopoulos and D. Kotsakis, *Cosmology* (Springer, Heidelberg, 1987).

[5] B. H. Bransden and C. J. Joachain, *Physics of atoms and molecules* (Longman, New York, 1983).

[6] D. N. Tripathy and S. Mishra, Int. J. Mod. Physics D, **7**, 431 (1998).

[7] A. L. Fetter and J. D. Walecka, *Quantum theory of many-particle systems* (McGraw-Hill, USA. 1971).

[8] J. M. Levy-Leblond, J. Math. Phys., **10**, 806 (1969).

[9] M. Karplus and R. N. Porter, *Atoms and Molecules* (Benjamin, Reading MA, 1970).

[10] E. R. Harrison, *Cosmology* (Cambridge University Press, Cambridge, England, 1981).

[11] S. L. Shapiro and S. A. Teukolsky, *Black Holes, White Dwarfs and Neutron Stars* (Wiley, New York, 1983).

[12] V. M. Lipunov, *Astrophysics of Neutron Stars* (Springer, Berlin 1992).

[13] C. E. Long, *Discovering the universe* (Harper and Row, New York, 1943).

[14] D. W. Sciama, *Modern Cosmology* (Cambridge, Cambridge University Press,1971).

[15] P. Flower, *Understanding the Universe* (West Publ. St. Paul, 1990).

Chapter 5

Black hole

In this chapter we study a black hole as another application of the many-particle Hamiltonian of chapter 3. Though we do not have a term in the Hamiltonian describing the curvature contribution, the variational method still gives very good analytical results due to self-consistency. Using a single particle density distribution for a system of self-gravitating particles which ultimately forms a black hole, we from a condensed matter point of view derive the Schwarzschild radius and by including the quantum mechanical exchange energy we find a small correction to the Schwarzschild radius, which we designate as the skin of the black hole.

5.1 Introduction

In the evolution of a star, when the radius of a star becomes less than a certain limit, called the Schwarzschild radius $R_{Sch} = 2GM/c^2$, where M is the mass of the object; G, Newton's universal gravitational constant, and c, velocity of light; neither light nor material particles can escape from the star. Thus, one finds from above, the Schwarzschild radius corresponds to the situation when the escape velocity is equal to the velocity of light, a particle (or a photon) coming from a large distance when passes near by the black hole [1, 2], it is not only attracted towards it but also its orbit diverges from a straight line. It may so happen that if the particle goes too close to a black hole it is likely to be trapped and hence, it cannot escape to infinity. As a special case, if the particle travels straight to the center of the black hole, it falls inside it and is lost forever. An interesting problem that is associated with the formation of a black hole is the final collapse of a massive star. This happens when the nuclear fuel inside the central core of the star gets exhausted. At this stage, it is the dominance of the gravitational attraction among the particles within a star over the internal outward pressure which makes the star to collapse. A black hole, though may be formed from baryons, leptons etc; the exterior observer cannot have access to the details of the inside of the black hole. The observer probes the black hole mass M, electromagnetic charge Q and angular momentum J. This is refereed to as baldness of the black hole or as Wheeler describes, a black hole has no hair [2]. But actually it has only three hairs, M, Q, J. We in this paper derive the Schwarzschild radius of a system of gravitating particles quantum mechanically and show that when quantum exchange correction is taken into account, there is a thin correction to this radius, which we designate as the skin of the black hole. It has been shown previously [3] that using the quantum field theoretic method, by including a single-closed-loop in the self energy, a quantum correction to the classical Schwarzschild solution of the order of $\sim G^2$ can be found. This comes from the gravity sector. The correction that we find, comes from the exchange part of the matter-energy sector of the black hole. Our correction is also of the order of $\sim G^2$.

We here consider a black hole with mass only, a Schwarzschild [1] type black hole. As described above, since the black hole forms out of particles due to immense gravitational attraction, we consider the system of particles as a many particle system that ultimately forms a black hole. We have succeeded in developing a quantum mechanical approach [4, 5] for a system of self-gravitating particles using Newtonian gravity. As done in the last chapter, we calculate the energy content of the system including exchange interaction. In order to calculate the total energy of a star, we choose a trial single-particle density to account for the distribution of particles within star. The form of our single-particle density is such that it has a singularity at the origin. Applying it to the case of a neutron star, we not only arrive at a compact expression for the radius of the neutron star, but also obtain an expression for the binding energy of the star which varies with the particle number [6] as $N^{7/3}$, where N is the particle number. Such a dependence with N is in agreement with those of the earlier workers. [6] The aim of the present work is to give a derivation of the so called Schwarzschild radius even without using GTR or relativistic quantum mechanics. By accounting for the exchange effects due to the interparticle correlations to the total ground state energy of the system, we find a quantum correction to the Schwarzschild radius.

5.2   Energy of a star

In order to describe a system of N self-gravitating particles in absence of any source for radiation, we use our standard Hamiltonian of the form:

$$H = \sum_{i=1}^{N} \frac{-\hbar^2 \nabla_i^2}{2m} + \frac{1}{2} \sum_{i=1}^{N} \sum_{j=1, i \neq j}^{N} v(|\vec{X}_i - \vec{X}_j|),$$ (5.1)

where $v(|\vec{X}_i - \vec{X}_j|) = -g^2/|\vec{X}_i - \vec{X}_j|$ is the interparticle interaction between pair of gravitating particles and $g^2 = Gm^2$, m being the mass of particle and G being the universal gravitational constant. As done in the last chapter, we calculate the total energy of the system and minimize it. Differentiating the energy with respect to λ and then equating it with zero, we obtain the value of λ at which the minimum occurs. This is found as: $\lambda_0 = \frac{72}{25}\left(\frac{3\pi}{16}\right)^{2/3}\frac{\hbar^2}{mg^2} \times \frac{1}{N^{1/3}}$. Here we are only concerned with the total kinetic energy of the system. At $\lambda = \lambda_0$, we have,

$$< KE >= 0.015441\frac{mg^4}{\hbar^2}N^{7/3} \tag{5.2}$$

The total energy E of the system is found to be just negative of this.

5.3 Schwarzschild radius: Quantum derivation

We now try to calculate the average velocity of a particle within the neutron star using Eq.(5.2). Let us denote it by $< \vec{v}^2 >$. If M denotes the total mass of the neutron star, one writes

$$< KE >= \frac{1}{2}M < \vec{v}^2 >, \tag{5.3}$$

where $M = Nm$. By comparing Eq.(5.3) with Eq.(5.2), one obtains,

$$< \vec{v}^2 >= 0.030882\frac{g^4}{\hbar^2}N^{4/3}. \tag{5.4}$$

From the expression for the total kinetic energy of an infinite many-fermion system [7], one finds that the average velocity of a particle within the system is $\sim 0.77v_f$, v_f being the fermi velocity [7] of the particle within the system, which is maximum velocity of that particle. From this, one clearly sees that the maximum velocity of a particle belonging to an infinite many-fermion system is greater than the average particle velocity. In view of this fact, we could write the maximum velocity of a particle within a neutron star as

$$< \vec{v}^2 >_{max}= \alpha < v^2 >= 0.030882\alpha\frac{g^4}{\hbar^2}N^{4/3} \tag{5.5}$$

where α is a constant whose value is to be greater than unity and it is to be calculated later. v_{max} can be identified as the escape velocity of a particle within

144

a neutron star. According to special theory of relativity, $\vec{v}_{max}^2$ is to be less than c^2, c being the velocity of light. That is,

$$0.030882\alpha\frac{g^4}{\hbar^2}N^{4/3} \leq c^2, \tag{5.6}$$

From this it follows that

$$N \leq \frac{13.574409}{\alpha^{3/4}}(\frac{\hbar c}{g^2})^{3/2} = N_c \ (say), \tag{5.7}$$

having $g^2 = Gm^2$. Substituting Eq.(5.7) in the expression for λ_0, one finds that,

$$\lambda_0 \geq \lambda_c = \frac{Gm}{c^2}\alpha^{1/4}[0.8483718(\frac{\hbar c}{g^2})^{3/2}]. \tag{5.8}$$

If we define the radius of a neutron star as $R_0 = 2\lambda_0$, we have the expression for the critical radius as,

$$R_c = 2\lambda_c = 2\frac{Gm}{c^2}\alpha^{1/4}[0.8483718(\frac{\hbar c}{g^2})^{3/2}] = \frac{2GM}{c^2} \tag{5.9}$$

Our identification about the radius R of the star with $2\lambda_0$ is based on the use of so called quantum mechanical tunneling [8] effect. Classically, it is well known that a particle has a turning point where the potential energy becomes equal to the total energy. Since the kinetic energy and therefore the velocity are equal to zero at such a point, the classical particle is expected to be turned around or reflected by the potential barrier. From the present theory it is seen that the turning point occurs at a distance $R = 2\lambda_0$. This is the reason why we identify $2\lambda_0$ with the radius of a star. For $R > 2\lambda_0$, a particle, belonging to the system, may have an access to the region beyond $R > 2\lambda_0$, because of quantum mechanical tunneling, but is forbidden by classical theory.

R_c as given in Eq.(5.9) is being identified as the so called Schwarzschild radius which we have derived here by treating the system as a quantum many-body system. When $R_0 \leq R_c$, the corresponding neutron star becomes a black hole. From Eq.(5.7), we therefore find that the lowest mass of the neutron star beyond which black hole formation takes place is given as

$$M_c = mN_c = \frac{13.574409}{\alpha^{3/4}} m \left(\frac{\hbar c}{g^2}\right)^{3/2} \tag{5.10}$$

In order to determine α, we now try to evaluate the limiting mass of a neutron star following the general expression for the radius of a star. Beyond this mass, the black hole formation is likely to take place. For that, we consider the situation when

$$(R_0 = 2\lambda_0) = \left(R_{sch} = \frac{2GM}{c^2}\right), \tag{5.11}$$

where $M = Nm$. From this, we arrive at

$$N \geq (1.696758) \left(\frac{\hbar c}{g^2}\right)^{3/2} = N_c. \tag{5.12}$$

Since the expression in the right hand side of Eq.(5.12) should be equal to the one given in right hand side of Eq.(5.7), we must have $\alpha = 16$. Under this situation, we have

$$v_{max}^2 = 0.494112 \frac{g^4}{\hbar^2} N_c^{4/3}, \tag{5.13}$$

where $N_c = 1.696758 \left(\frac{\hbar c}{g^2}\right)^{3/2}$, which, when evaluated, becomes 3.7390777×10^{57}. For a neutron star in which the number of neutrons exceeds N_c, it has the tendency of forming a black hole. In that case, its mass must exceed $M = M_c = mN_c = 3.12213\ M_\odot$, $M_\odot$ being the solar mass.

5.4   Quantum correction to Schwarzschild radius

So far we have been discussing about the quantum mechanical derivation of the Schwarzschild radius R_{Sch}. The very form of R_{Sch} shows that it is a classical result, leaving aside the fact the number of particles N within a neutron star [9] is to be less than N_c where $N_c = 1.70 \left(\frac{\hbar c}{Gm_n^2}\right)^{3/2}$, which involves the Planck's constant $\hbar$. Now, in order to account for the quantum corrections to R_{Sch}, we go beyond the Hartree contribution to the total energy of the system. That is the exchange correction or Hartree-Fock(HF) term [10] over the Hartree result (direct contribution). Since the HF-correction term is non-local we make use of the local density approximation [10] to write it as,

$$< PE >_{ex} = \frac{3}{2\pi} (3\pi^2)^{1/3} g^2 \int d\vec{r} [\rho(\vec{r})]^{4/3}. \tag{5.14}$$

This when evaluated (using the trial density $\rho(r)$ of last chapter) gives

$$< PE >_{ex} = \frac{27}{4} (\frac{1}{16\pi})^{4/3} (3\pi^2)^{1/3} g^2 \frac{N^{4/3}}{\lambda}. \tag{5.15}$$

With the inclusion of this extra term in the expression for total energy of the last chapter, we minimize it with respect to λ and arrive at

$$\lambda_0' = \frac{72}{25} (\frac{3\pi}{16})^{2/3} \frac{\hbar^2}{mg^2} \frac{1}{N^{1/3}} [1 + \frac{1.8010}{N^{2/3}}] \tag{5.16}$$

Following the argument discussed earlier, we identify the radius of the neutron star by $R_0' = 2\lambda_0'$. As before writing $v'^2_{max} = 16 < v'^2 >$ and using the condition that $v'^2_{max} \leq c^2$, we obtain

$$N \leq N_c' = 1.696758 (\frac{\hbar c}{g^2})^{3/2} [1 + 0.4747761 (\frac{g^2}{\hbar c})] \tag{5.17}$$

Corresponding to N_c', the new expression for the critical radius R_c' becomes

$$R_c' = 2\lambda_c' = 2 \frac{GM_c}{c^2} [1 + 0.7912723 (\frac{g^2}{\hbar c})] \tag{5.18}$$

where $M_c = mN_c = 1.696758 (\frac{\hbar c}{g^2})^{3/2}$. The above expression, Eq. (5.18) is obtained by keeping terms upto order $(\frac{g^2}{\hbar c})$ only in Eq. (5.16). For $N > N_c'$, the neutron star is likely to go over the black hole stage. From Eq.(5.18), we find that the second term within the square bracket, forms the quantum correction to the Schwarzschild radius. As expected, it involves the gravitational fine structure constant $(\frac{g^2}{\hbar c})$. Since it is of the order 10^{-39}, obviously it makes an extremely small correction to R_{Sch}. It has been shown earlier [3] that using the quantum field theoretic method and by including a single-closed-loop in the self energy, a quantum correction to the classical Schwarzschild solution of the order of $\sim G^2$ can be found. That comes from the gravity sector. The correction that we get is also of the order $\sim G^2$ but it comes from the exchange part of the matter-energy sector of the black hole.

5.5 Summary

We in this chapter derive the Schwarzschild radius of a black hole from a condensed matter point of view by using a single particle density distribution for the many-body self-gravitating system which ultimately forms a black hole. By incorporating the quantum mechanical exchange interaction, we also find a small correction to the Schwarzschild radius which we designate as the skin of the black hole.

Bibliography

[1] D. Raine and E. Thomas, *Black Holes* (Imperial College Press, London, 2005).

[2] J. A. Wheeler and K. Ford, *Geons, black holes and quantum foam: A life in physics* (Norton, 2000).

[3] M. J. Duff, Phys. Rev. D. **9**, 1837 (1974).

[4] D. N. Tripathy and S. Mishra, Int. J. Mod. Phys. D**7**, 431 (1998).

[5] D. N. Tripathy and S. Mishra, Int. J. Mod. Phys. D**7**, 917 (1998).

[6] J. M. Levy-Leblond, J. Math. Phys. **10**, 806 (1969).

[7] A. L. Fetter and J. D. Walecka, *Quantum theory of many-particle system* (McGraw Hill, New York, 1971).

[8] M. Karplus and R. N. Porter, *Atoms and Molecules* (Benjamin, California, 1970).

[9] P. S. Wesson, *Cosmology and Geophysics* (Adam Hilger, Bristol, 1978).

[10] H. A. Bethe and R. W. Jackiw, *Intermediate Quantum Mechanics* (Benjamin, London, 1968).

Chapter 6

The Universe

In this chapter we relativize the Hamiltonian of Eq.(2.204) of the quantized Newton-Cartan-Schrodinger system to study the static properties of our universe. Since after inflation, 10^{-35} s after the Big Bang of the universe the spatial curvature became almost zero, we consider the case when spatial curvature $k = 0$.

6.1   Facts about our universe

The current theory for the origin of the universe is the Big Bang theory [1, 2] according to which the universe is considered to have started with a huge explosion from a superdense and superhot state. Theoretically, it means that the universe has started from a mathematical singularity with infinite density. the expansion of the universe is being described by the Hubble's law [3] which is given by the relation

$$v = H_0 d \qquad (6.1)$$

where v denotes the radial velocity at which each distant galaxy is receding from us, d is the distance of the galaxy and H_0 is the Hubble constant. The time which would have elapsed since the scale factor $R = 0$ to the present stage of the universe is given by $t_0 = 1/H_0$, where t_0 is called the Hubble time. The Hubble time represents the maximum age of the universe.

In spite of the fact that the Big Bang theory proves to be very successful description of the universe for the whole range of times starting almost from Big Bang to the present age, it has a few serious shortcomings in the sense that using this theory, a number of very obvious questions have been left yet unanswered. The first question involves the ratio of the number of protons and neutrons in the universe to the number of photons. Photons are found mainly in the cosmic back ground radiation, while protons and neutrons form the atomic nuclei of matter that make up the galaxies. From the calculated abundances of the nuclides mentioned earlier, it is concluded [4] that the observed universe contains $\sim 10^9$ photons for every proton or neutron. The standard Big Bang theory (SBBT) does not explain this ratio but instead assumes that the ratio is given as a property of the initial condition. There has been an estimate of the average density of photons in the universe considering the fact that most of the diffuse photons form the microwave background radiation [5–7]. From this it follows that there should be about 500 photons per cubic centimeter of the universe. It has also been established that there would be an equal number of neutrinos too, per cubic centimeter in the universe. This has not been explained using SBBT. The third question which has not been answered yet by the SBBT is related to the large scale homogeneity of the observed universe. Although the universe we observe is, in many ways, quite inhomogeneous because of the presence of stars, galaxies and cluster of galaxies within it, when looked over very large scales, it apperas to be very homogeneous. The last thing for which there is no explanation in the SBBT is related to the mass density ρ_0 of the universe, which is measured relative to some critical density denoted by ρ_c. If ρ_0 exceeds ρ_c, then the gravitational pull of everything over everything else will be strong enough to halt the expansion, eventually without considering accelerated expansion. This would cause the universe to collapse, resulting in what is sometimes called a Big crunch. If on the otherhand, $\rho_0 < \rho_c$, the universe will go on expanding forever. Cosmologists introduce a quantity, designated by a symbol Ω, which is defined by the ratio of the mass density to the critical mass density (ρ_0/ρ_c). The correct value is $\Omega \approx 1$, where $\Omega = \Omega_m + \Omega_\Lambda$ and $\Omega_m = \rho/\rho_c$, $\Omega_\Lambda = \rho_\Lambda/\rho_c$ and the critical density $\rho_c = 3H^2/8\pi G$.

Assuming the best value of the Hubble constant to be equal to $\approx 70 \ km/s$ $/Mpc$, which corresponds to an age of the universe as $t = 13.7 \ Gyr$, we calculate [3] ρ_c which gives $\rho_c = 6.8 \times 10^{-30} \ g/cm^3$. Measurement of the average density of universe ρ_0 is extremely difficult, because the universe consists of all sorts of matter and all of them contribute to ρ_0. Actually now it is known that 70% of total mass comes from the vacuum energy or cosmological constant and baryonic matter amounts to only 5% and the nonbaryonic dark matter constitutes 25% of the critical density of the universe, that is $\rho_{DM} = 1.7 \times 10^{-30} \ g/cm^3$. The neutrinos which are hot or relativistic constitute only 1% of the dark matter. The large scale structure formation implies that most of the dark matter should be cold or nonrelativistic. However, the average density of matter in a galaxy can be determined using the relation $\rho_G = n_G M_G$, where M_G is the mean mass of a galaxy and n_G, the number of galaxies in a unit volume. For some fixed region of the sky, we count the number of galaxies and find n_G. In order to determine M_G, we consider the fact that the stars in a galaxy revolve around the center of galaxy. This is equivalent to the motion of planets around the sun. Hence once we know the rotating velocity of some of the stars in a galaxy and distances from the galactic center, M_G can be found out [4]. Such measurements show that $\rho_G \approx 3 \times 10^{-31} \ g/cm^3$. assuming that the cosmic matter is mainly in galaxies we have $\rho_0 \approx \rho_G$. This means that $\rho_0 < \rho_c$ and hence, the universe will be infinite, open and ever expanding.

The very assumption that matter in the universe is mainly concentrated in galaxies need not be correct. It is very likely that the space between the galaxies is not a vacuum but contains gases or extinguished stars [8]. The question is, just how much of such unknown matter is there in the universe? Looking at the work of Zwicky [9] relating to the measurement of the mass of coma cluster, we find that the dynamical mass of coma cluster is almost four hundred times its luminosity mass. The luminosity mass of a cluster is determined by measuring the luminosities of the member galaxies of a cluster and then adding up all the masses of the members. As regards the dynamical mass is concerned, this is determined by measuring the relative velocities among the galaxies. In this case, the mean relative velocity is related to the whole cluster. The great discrepancy one notices in the two measured values for the masses of the coma cluster can only be interpreted by the fact that the mean mass of the coma cluster is not only contributed by the visible galaxies, but by a large amount of invisible matter within the cluster. The mass measured from the luminosities includes only the mass in the light emitting regions and does not include the mass that exists in regions not emitting light. If non light emitting region contains a large amount of matter, then only the luminosity mass would be much less than dynamical mass. Thus, from Zwicky's work, we get an idea about the missing mass or about the Dark Matter (DM) of the present universe.

Since light emitting bodies must have a lot of baryons and the dominant contribution to the cosmic mass comes from the DM, it is therefore believed that the particles constituting the DM can not be baryons. Besides, particles of DM are expected not to participate in electromagnetic interactions. If they are governed by weak interactions, then only they can not be detected in today's laboratory experiments. Out of the list of non-baryonic kind of particles that have been suggested to constitute the DM of the universe, the neutrinos which are hot or relativistic constitute only 1% of the dark matter. Other dark matter particle candidates are weakly interacting massive particles (WIMP) [10] which may be massive with mass in order of GeV considered to be the most suitable candidates to form the particles for the DM. The neutrinos have a non-zero rest mass. A recent study [11] by Schramm and Steigmann suggests that mass of the neutrino must be between (4 to 20 eV/c^2). Actually the recent [12] upper bound on the neutrino mass is $m_\nu = 0.20 \ eV$ and the lower bound is $m_\nu > 10^{-3} \ eV$.

Though the cosmological constant can contribute 70% to the critical density, the recent realization is that since the scale of cosmological constant and neutrino masses are of the same order of magnitude, some kind of radiative models of neutrino masses which require Higgs scalar, could be a dark matter candidate. Though the present upper limit on neutrino mass does not permit standard neutrinos to be dark matter candidates, right-handed sterile neutrinos could be a dark matter candidate which again depend on their mass mixing with active neutrinos [13]. Considering the fact that the major constituents of the present universe are dark matter particle, which might have been formed out of some fictitious particles which represent effectively the DM and Cosmological constant contribution, and these particles are of certain mass that had filled the early universe at a time some two minutes after the Big Bang, we, in this chapter, have tried to calculate the mass, mean density and radius of the universe by treating the universe just as a system made out of these particles which are self gravitating. The present calculation is based on making an intuitive choice for the distribution of particles in the universe, characterized by a distribution function having a singularity at the origin. Such a form of single particle density seems to be consistent with the concept like the Big Bang theory of the Universe. The present calculation is the result of our earlier study [14–16] where such a form of singular density distribution has been used to calculate the binding energy of a system of self gravitating particles like the neutron stars and white dwarfs etc., having no source for nuclear power generation present at their cores. In all these cases our theory has proved to be of a great success due to the fact that it has correctly reproduced the results for the binding energies and the radii of the neutron stars etc, that agree with those obtained by other workers. In the present work, we assume the fictitious particles to be fermionic in nature, carrying a tiny mass m, which interact among themselves through gravitational forces. Considering the fact that the present age [4] of the universe is about 14×10^9 yr, we have been able to make an estimate for the radius of the universe by adjusting the mass of these fictitious particles. The age of the Universe is determined using the fact that the horizon of the universe has been expanding with a speed equal to the speed of light $'c'$. With this, we not only arrive of a value of $\sim 10^{28}$ cm for the radius of the universe but also we obtain a value of its mass of about $\sim 10^{23} M_\odot$, $M_\odot$ being the solar mass ($M_\odot = 2 \times 10^{33} g$). All these numbers seem to be matching very nicely with the corresponding results known from other theories. Our estimated result for the average mass density of the present universe comes out to be $\approx 10^{-29} g/cm^3$, which is supposed to be the case as per expectation. In sec.2 of this chapter the mathematical formalism for the present calculation is presented. In sec.3 Mach's principle is introduced to relativize the problem. In sec.4 the results are given. Our calculated value for the mass of the particles responsible for the formation of the Dark Matter seems to be agreeing with the mass of neutrinos as speculated by several other workers and in this section, we have also presented our estimated values for the ratio of the variation of the universal gravitational constant G with time to G itself, that is ($\frac{\dot{G}}{G}$). This is again found to be in extremely good agreement with those obtained from some of the most recent calculations [6].

6.2 The quantum theory of the universe

If we visualize the present universe to be a system composed of some self-gravitating fictitious particles, each of mass m, interacting through pair-wise gravitational interactions, then the Hamiltonian of the system can be written as

$$H = -\sum_{i=1}^{N}(\frac{\hbar^2}{2m})\nabla_i^2 + \frac{1}{2}\sum_{i=1,i\neq j}^{N}\sum_{j=1}^{N}v(|\vec{X}_i - \vec{X}_j|), \qquad (6.2)$$

where $v(|\vec{X}_i - \vec{X}_j|) = -g^2/|\vec{X}_i - \vec{X}_j|$, having $g^2 = Gm^2$, G being Newton's universal gravitational constant. As discussed in chapter 1 and 2, recently it is shown that the Newton-Cartan limit (that is $c \to \infty$ and the 3-space being flat though the spacetime is still curved) of Einstein' General Theory of Relativity can be exactly quantized [17] without any socalled time problem and it leads to the above Hamiltonian Eq.(6.2).

The effective particles which constitute the universe are treated as fermions. In our calculation, we use the zero temperature formalism which can be justified on the ground that the present cosmic background temperature, being 2.728K, is close to 0K. It is well known that the temperature in the early universe was extremely high. In order to simulate the condition in the early universe, using the present theory, we have to assume that the present total kinetic energy of the system was then available in the form of an equivalent temperature T by virtue of the relation $<KE> = (\frac{3}{2})Nk_BT$, the factor $(\frac{3}{2})Nk_B$ comes from all the three degrees of freedom of a particle.

As done in the chapter 3, in the Thomas-Fermi approximation [14], we evaluate the total kinetic energy of the system using the relation :

$$<KE> = \left(\frac{3\hbar^2}{10m}\right)(3\pi^2)^{2/3}\int d\vec{X}[\rho(\vec{X})]^{5/3}, \qquad (6.3)$$

where $\rho(\vec{X})$ is the single particle distribution function which the constituent

156

particles obey within the universe, such that

$$\int d\vec{X} \rho(\vec{X}) = N, \tag{6.4}$$

N being the total number of these particles in the universe. The total potential energy of the system in the Hartree-approximation is obtained as

$$< PE > = -(\frac{g^2}{2}) \int d\vec{X} d\vec{X'} \frac{1}{|\vec{X} - \vec{X'}|} \rho(\vec{X}) \rho(\vec{X'}) \tag{6.5}$$

Evaluation of the integrals shown in Eq.(6.3) and Eq.(6.5) is done using a single particle density distribution of the form:

$$\rho(x) = \frac{Ae^{-x}}{x^3}, \tag{6.6}$$

where $x = (r/\lambda)^{1/2}$, $r = |\vec{X}|$. Such a form of the single particle density for the ground state has been used by us [14–16] while calculating the binding energy of a neutron star and its radius. After evaluating the integrals given in Eq.(6.3) and Eq.(6.5), the expression for total energy of the universe, $E(\lambda)$, becomes

$$E(\lambda) = \left(\frac{12}{25\pi}\right)\left(\frac{\hbar^2}{m}\right)\left(\frac{3\pi N}{16}\right)^{5/3}\frac{1}{\lambda^2} - \left(\frac{g^2 N^2}{16}\right)\frac{1}{\lambda} \tag{6.7}$$

Now, minimizing $E(\lambda)$ with respect the λ and then evaluating it at $\lambda = \lambda_0$, where λ_0 is the value of λ at which the minimum occurs, the total binding energy of the universe, corresponding to its lowest energy state, becomes

$$E_0 = -(0.015442)N^{7/3}(\frac{mg^4}{\hbar^2}) \tag{6.8}$$

In view of our earlier works [14, 15] and as discussed in the previous chapter, here also we identify $2\lambda_0$ result with the radius R_0 of the universe. It must be noted that the size of any compact object (either an atom or a star) is not well defined in quantum theory. The justification regarding the identification of the radius R_0 of a star with $2\lambda_0$ follows from the consideration of the so called tunneling effects used in quantum mechanics. Classically,it is known that a particle has a turning point where the potential energy becomes equal to the total energy [18, 19]. Since the kinetic energy and therefore the velocity are equal to zero at such a point,the classical particle is expected to be turned around or reflected by the potential barrier. For example,considering the case of an electron in the hydrogen atom ground state such classical turning point occurs where the potential $V(r) = -e^2/r = E_{total} = -e^2/2a_0$; that is at $r = 2a_0$. Quantum mechanically,the probability distribution $r^2\rho(r)$ has a non-zero value for $r > 2a_0$; that is, the electron has access to the region $r > 2a_0$ which is forbidden by classical theory. Such penetration or tunneling into or through the potential energy barriers is typical of results of quantum theory. If the electron had a value of $r > 2a_0$, then its kinetic energy would have to be negative to satisfy the condition $E_{total} = T + V$,with $V > E_{total}$. Since negative kinetic energy is physically absurd, $r = 2a_0$ is to be identified as the classical radius. Using the above idea, from the present theory one can easily see that at $\lambda = 2\lambda_0$, the potential energy of the system becomes equal to the the total energy, there by proving that the radius of the system $R_0 = 2\lambda_0$.

$$R_0 = 2\lambda_0 = (\frac{\hbar^2}{mg^2}) \times (4.047528)/N^{1/3} \qquad (6.9)$$

6.3   Mach's principle

We now invoke Mach's principle [20] which states that inertial properties of matter are determined by the distribution of matter in the rest of the universe. Mach had the view [1] that the velocity and acceleration of a particle would be meaningless had the particle been alone in the universe. We have to talk of acceleration only with respect to other bodies, just like we talk of velocities with respect to other bodies. This means that the inertial mass of a particle is the result of the particle feeling the presence of other particles in the universe. If we denote the inertial mass of the particle by m_{inert}, it is to be determined by its response to accelerated motion. As far as the universe is concerned, the distance particles beyond the Hubble length, which we take as the radius of the visible universe R_0 are unobservable and therefore do not contribute to the determination of local inertial mass. If M denotes the gravitational mass of the observable universe, the gravitational energy of the particle is given by $E_{gr} = \frac{GMm_{grav}}{R_0}$, where m_{grav} is the gravitational mass of the particle, that is, the mass determined by its response to gravity. In accordance with the spirit of Mach's principle, one must have $E_{gr} = \frac{GMm_{grav}}{R_0} = m_{inert}c^2$, where $m_{inert}c^2$ is the intrinsic energy of the particle. Since m_{inert} and m_{grav} are taken to be equal both in Newtonian theory and in the General Theory of Relativity, we have Mach's principle [20] expressed through the relation as

$$\frac{GM}{R_0 c^2} \approx 1 \qquad (6.10)$$

and using the fact that the total mass of the universe $M = Nm$, we are able to obtain the total number of particles N constituting the universe.

6.4 Results

Using Eq.(6.10) in the expression $R_0 = 2\lambda_0$, Eq.(6.9) one arrives at

$$N = 2.8535954\left(\frac{\hbar c}{Gm^2}\right)^{3/2} \qquad (6.11)$$

Substituting the expression for N from Eq.(6.11) in Eq.(6.9), the expression for R_0 becomes

$$R_0 = 2.8535954\left(\frac{\hbar}{mc}\right)\left(\frac{\hbar c}{Gm^2}\right)^{1/2} \qquad (6.12)$$

TABLE - I

$m \times 10^{-35} g$	$R \times 10^{28} cm$	$N \times 10^{91}$	$M \times 10^{56} gm$	$\tau_0 \times 10^9 yr$
1.07299	1.896	2.38429	2.55832	20
1.23891	1.422	1.54865	1.91875	15
1.51744	0.948	0.84297	1.27916	10
2.14598	0.474	0.29804	0.639588	5

Using Eq.(6.12), we have made an estimation for the the radius of the present universe, R_0, by varying m, the mass of the fictitious particle and using the measured value for the gravitational constant G, that is by taking $G = 6.67 \times 10^{-8}$ $dyn\ cm^2\ g^{-2}$. We have chosen a set of four values for m and calculate R_0, from which the age of the Universe τ_0 is estimated using the relation $R_0 \simeq c\tau_0$. All these are shown in Table-I. Accepting $\tau_0 = 15 \times 10^9\ yr$, to be the most correct value for the age, we find that this corresponds to a value of $R_0 \simeq 1.422 \times 10^{28}\ cm$. This is obtained from Eq.(6.12) by choosing $m = 1.23891 \times 10^{-35}\ g$. For this m, we calculate N following Eq.(6.11), which gives $N \sim 10^{91}$. This gives rise to the total mass of the universe as $M \simeq Nm \sim 10^{23} M_\odot$, $M_\odot$ being the solar mass, whose value is $2 \times 10^{33}\ g$.

Using the results for M and R of the present universe of age $\approx 15 \times 10^9\ yr$ from Table-I, nucleon number density, the ratio of number of nucleons to number of photons etc. are calculated which are given in Table-II.

We now calculate the variation of the gravitational constant G with respect to time, $\dot{G}(t)$. For that, we, use the expression for G, as found from Eq.(6.12), which is

$$G = K/R_0^2, \qquad (6.13)$$

where

$$K = (8.1430067)(\frac{\hbar^3}{m^4 c}) \qquad (6.14)$$

Further ,we assume that the above expression for G is also valid for anytime 't'

TABLE - II

$M_n \times 10^{56}g$	$N_n \times 10^{78}$	$N_\nu/N_n \times 10^9$	$N_\nu \times 10^{88}$	N_ν/V	$M_\nu \times 10^{56}g$	$m_\nu \times 10^{-32}g$
0.03 (1.91875)	3.4406	3.344	1.15	950	1.8612	1.6176
0.03 (1.91875)	3.4406	3.125	1.08	890	1.8612	1.7309
0.03 (1.91875)	3.4406	1.740	0.60	500	1.8612	3.0900
0.03 (1.91875)	3.4406	1.000	0.34	285	1.8612	5.4090

provided we take the value of R_0 at that time. Using Eq.(6.13), we therefore, find

$$\dot{G} = \left(\frac{\partial G}{\partial R_0}\right) \times \left(\frac{\partial R_0}{\partial t}\right) \simeq c\left(\frac{\partial G}{\partial R_0}\right) = \frac{-2cK}{R_0^3} \tag{6.15}$$

Following this, we make an estimation of $(\frac{\dot{G}}{G})$, for $m = 1.24 \times 10^{-35}$ g which gives

$$\left(\frac{\dot{G}}{G}\right) = -1.3 \times 10^{-10} yr^{-1} \tag{6.16}$$

The above value is in extremely good agreement with the recent estimates [6]. This is roughly also the value reported by Van Flandern [22] following an analysis of the data relating to the effects of tidal friction on the elements and shape of the lunar orbit of the earth-moon system. However Shapiro et al [23], have arrived at a limiting value $| (\frac{\dot{G}}{G}) | \leq 4 \times 10^{-10} yr^{-1}$, based on a method which is used for monitoring the planets for a possible secular increase in their orbital periods by employing a radar reflection system between the Earth,Venus and Mercury.

6.5   Discussion of results

In arriving at the various results of the present theory, we have accepted the value for the age of the universe τ_0 to be $\sim 15 \times 10^9 yr$. The mass of the constituent particles corresponding to this is found to be $m \sim 1.24 \times 10^{-35}g$. It is also to be noted that in the present theory we have treated the constituent particles as fermions. For a mass of $m \sim 1.24 \times 10^{-35}g$, the radius of the universe becomes $R_0 \sim 1.422 \times 10^{28}$ cm, which is considered to be an acceptable result, and total mass of the universe M_0 becomes $M_0 \sim (1 \times 10^{23})M_\odot$, $M_\odot$ being the solar mass. If we assume the mass 3% of M to be solely due to the nucleons that are there in the today's universe, it would correspond to $\sim 10^{78}$ nucleons to be present. This is nothing but the famous Eddington result [20, 21]. Let us now look at our calculated value for the average mass density of the universe. From Table-I we see that, corresponding to $m \sim 1.23891 \times 10^{-35}g$ the average mass density of the universe is $\sim 10.6 \times 10^{-30}$ g/cm^3. This, we find to be almost thirty times larger then the observed mass density, ρ_0, of the present universe, $\rho_0 \sim 3 \times 10^{-31}g/cm^3$ where the latter is being thought to be arising solely due to nucleons and other heavy elements etc, present. Let us now assume that out of the entire mass M_0 of the universe, some percentage of it has gone into the formation of the nucleons and other heavy elements present today and the remaining part has been lying in the form of Dark Matter in the present universe. This amounts to saying that the total number of fictitious particles responsible in the formation of the early universe have been subsequently converted to appear in the present universe as nucleons and other heavy nuclei and some other form of particles contributing to the Dark Matter of the universe. Let us accept the view that not more than three percent of the entire mass of the universe is contributed by the nucleons and other heavy nuclei [9], the contribution from the heavy nuclei being negligibly small compared to that of the nucleons. This would therefore mean that the remaining ninety-seven percent of its total mass constitutes what is known as the Dark Matter and Dark energy of the universe. Having accepted this picture, we have calculated the ratio of the number of neutrinos to nucleons, (N_ν/N_n), and the neutrino number per unit volume of the universe, (N_ν/V), etc, V being the volume of the universe.

From the above table-II we find that the calculation done by assuming that only three percent of the total mass of the universe constitutes the observed density of the universe, is the most appropriate one. Because, for this case, the average density of the nucleons (which constitute the observable matter of the universe) comes out to be $\rho_0 \sim 4.8 \times 10^{-31}$ g/cm^3, which is considered close to the accepted value. This corresponds to an age of 15 billion years for the universe, and for this the total nucleon number of the universe becomes $\sim 3.5 \times 10^{78}$. This being so, the remaining ninety seven percent of the total mass of the universe is considered to be solely due to the Dark Matter and Dark Energy present within it. Accepting the fact that this mass is generated by the neutrinos present in the universe, we have made an estimation of the ratio (N_ν/N_n) and (N_ν/V) by varying the mass of the neutrinos. From table-II, one can see that for a neutrino mass of $m_\nu \sim 3.09 \times 10^{-32}$ g, one arrives at $(N_\nu/N_n) \simeq 1.74 \times 10^9$ and $(\frac{N_\nu}{V}) \simeq 500$. A mass of $m_\nu = 3.09 \times 10^{-32} g$ corresponds to an energy of 14 eV, which is again found to be the right order of magnitude for the upper limit of the neutrino mass.

Calculating the value for the average mass density of the present universe (Table-II) which comes out to be $\rho_U \simeq 10.6 \times 10^{-30} g/cm^3$, one can clearly see that it is obviously much larger than the observed mass density ρ_0 ($\rho_0 \simeq 4.8 \times 10^{-31} g/cm^3$). But it is interesting to note that the total mass density including the cosmological constant contribution (which is equivalently included here through the fictitious particles) is of the correct order of magnitude and equal to the critical density $\rho_U = (\rho_c \sim 6.8 \times 10^{-30}$ $g/cm^3)$. In the next chapter we will include the contribution due to cosmological constant explicitly.

To conclude, we have shown that the results of the present calculation are obtained by choosing a singular form of single particle density for the particles constituting the universe and a singular density is consistent with the idea relating to the Big Bang theory. Since the standard Big Bang theory so far has not succeeded to explain for the ratio of $\approx 10^9$ for (number of photons/number of nuclei) and here we do reproduce the above number correctly, the present work seems to be justifying the so called Big Bang theory of the universe.

6.6   Summary

Considering the fact that the present universe might have been formed out of a system of fictitious self-gravitating particles, fermionic in nature, each of mass m, we are able to obtain a compact expression for the radius R_0 of the universe by using a model density distribution $\rho(r)$ for the particles which is singular at the origin. This singularity in $\rho(r)$ can be considered to be consistent with the so called Big Bang theory of the universe. By assuming that Mach's principle holds good in the evolution of the universe, we determine the number of particles, N, of the universe and its R_0, which are obtained in terms of the mass m of the constituent particles and the Universal Gravitational constant G only. It is seen that for a mass of the constituent particles $m \simeq 1.24 \times 10^{-35}g$, the age of the present universe, τ_0, becomes $\tau_0 \simeq 15 \times 10^9 yr$, or equivalently $R_0 \simeq 1.4 \times 10^{28} cm$. For this m, the total number of particles constituting the present universe is found to be $N \simeq 1.55 \times 10^{91}$ and its total mass $(M \simeq 1 \times 10^{23} M_\odot)$, $M_\odot$ being the solar mass. All these numbers seem to be quantitatively agreeing with those evaluated from other theories. Using the present theory, we have also made an estimation of the variation of the universal gravitational constant G with time which gives $(\frac{\dot{G}}{G}) = -1.3 \times 10^{-10} \ yr^{-1}$. This is again in extremely good agreement with the results of some of the most recent calculations. Lastly, a plausible explanation for the Dark Matter present in today's universe is given. Assuming neutrinos to be one of the most possible candidate for the Dark Matter, we have estimated the ratio of the number of neutrinos to nucleons and the number of neutrinos per unit volume of the universe, which respectively gives $(\frac{N_\nu}{N_n}) \sim 1.74 \times 10^9$, $(\frac{N_\nu}{V}) \sim 500$. Both these numbers seem to be in agreement with the findings of the recent observations. The present calculation gives a mass for the neutrino to be $m_\nu \sim 3.09 \times 10^{-32} g$ or equivalently, $14\ eV$, which is the right order of magnitude, as speculated by several workers If we consider that only 1% of the total mass of the universe is due to neutrinos, then mass of a neutrino can be estimated to be $(14/97)\ eV \approx 0.14\ eV$ which is in very good agreement with the limits set for its mass.

Bibliography

[1] G. Contopoulos and D. Kotsakis, *Cosmology* (Springer, Heidelberg, 1987).

[2] A. H. Guth, in *Bubbles, voids and bumps in time: the new cosmology* edited by. J. Cornell (Cambridge University Press, Cambridge, 1989).

[3] U. Sarkar, *Particle and Astroparticle Physics* (Taylor & Francis, New York, 2007).

[4] F. L. Zhi and L. S. Xian, *Creation of the Universe* (World Scientific, Singapore, 1989).

[5] A. A. Penzias and R. W. Wilson, Astrophys. Jour. **142**, 419 (1965).

[6] J. V. Narlikar, *Introduction To Cosmology* (Cambridge University Press, London, 1993).

[7] G. S. Kutter, *The universe and life* (Jones and Bartlett, USA, 1987).

[8] I. Novikov, *Black holes and the Universe* (Cambridge University Press, Cambridge, 1990).

[9] F. Zwicky, Helve. Phys. Act, **6**, 10 (1933).

[10] J. E. Gunn in *Bubbles,voids and bumps in time: the new cosmology* edited by J. Cornel (Cambridge University Press, Cambridge, 1989).

[11] D. N. Schramm, Phys. Today, **36**, 27 (1983); G. Steigman, Ann. Rev. Astron. Astrophys. **14**, 339 (1976).

[12] W. Buchmuller, P. Di Bari and M. Plumacher, Phys. Lett. **B 547**, 128 (2002).

[13] R. Fardon, A. E. Nelson and N. Weiner, JCAP, **10**, 005 (2004).

[14] D. N. Tripathy and S. Mishra, Int. J. Mod. Phys. D **7**, 6, 917 (1998).

[15] D.N.Tripathy and S. Mishra Int. J. Mod. Phys. D **7**, 3, 431 (1998).

[16] S. Mishra, Int. J. Theor. Phys. **47**, 2655 (2008).

[17] J. Christian, Phys. Rev. D, 56, 4844 (1997).

[18] M. Karplus and R. N. Porter, *Atoms and Molecules* (Benjamin, Reading MA 1970).

[19] L. D. Landau and E. M. Lifshitz, *Quantum Mechanics* (Pergamon Press, Oxford, 1965).

[20] E. R. Harrison, *Cosmology* (Cambridge University Press, Cambridge, 1981).

[21] P. S. Wesson, *Cosmology and Geophysics* (Adam Hilger Bristol,1978).

[22] T. C. Van Flandern, *Mon. Not. Roy. Astron. Soc.*, **170**, 333 (1975).

[23] I. I. Shapiro, BAAS, **8**, 308 (1976).

[24] C. Brans and R. H. Dicke, Phys. Rev. **124**, 125 (1961).

Chapter 7

The Expanding Universe

Considering our expanding universe as made up of gravitationally interacting particles which describe particles of luminous matter, dark matter and dark energy which is described by a repulsive harmonic potential among the points in the flat 3-space, we derive a quantum mechanical relation connecting, temperature of the cosmic microwave background radiation, age, and cosmological constant of the universe. When the cosmological constant is zero, we get back Gamow's relation with a much better coefficient. Otherwise, our theory predicts a value of the cosmological constant 2.0×10^{-56} cm^{-2} when the present values of cosmic microwave background temperature of 2.728 K and age of the universe 14 billion years are taken as input.

7.1 Cosmological constant

The most important theory for the origin of the universe is the Big Bang Theory [1] according to which the present universe is considered to have started with a huge explosion from a superhot and a superdense stage. Theoretically one may visualize its starting from a mathematical singularity with infinite density. This also comes from the solutions of the type I and type II form of Einstein's field equations [2]. What follows from all these solutions is that the universe has originated from a point where the scale factor R (to be identified as the radius of the universe) is zero at time $t = 0$, and its derivative with time is taken to be infinite at this time. That is, it is thought that the initial explosion had happened with infinite velocity, although, it is impossible for us to picture the initial moment of the creation of the universe. The accelerated expansion of the universe has been conformed by studying the distances to supernovae of type Ia [3, 4]. For the universe, it is being said that the major constituent of the total mass of the present universe is made of the Dark Energy 70%, Dark Matter about 26% and luminous matter 4%. The Dark energy is responsible for the accelerated expansion of the universe since it has negative pressure and produces repulsive gravity. The cosmological constant [5, 6] of Einstein provides a repulsive force when its value is positive. The cosmological constant is also associated with the vacuum energy density [7] of the space-time. The vacuum has the lowest energy of any state, but there is no reason in principle for that ground state energy to be zero. There are many different contributions [7] to the ground state energy such as potential energy of scalar fields, vacuum fluctuations as well as of the cosmological constant. The individual contributions can be very large but current observation suggests that the various contributions, large in magnitude but different in sign delicately cancel to yield an extraordinarily small final result. The conventionally defined cosmological constant Λ is proportional to the vacuum energy density ρ_Λ as $\Lambda = (8\pi G/c^2)\rho_\Lambda$. Hence one can guess that $\rho_\Lambda = \Lambda c^2/8\pi G \approx \rho_{Pl} = c^5/G^2\hbar \sim 5 \times 10^{93}$ g cm^{-3}, where ρ_{Pl} is the Plank density. But the recent observations of the luminosities of high redshift supernovae gives the dimensionless density $\Omega_\Lambda = \rho_\Lambda/\rho_{cr} \equiv \Lambda c^2/3H_0^2 \approx 0.7$ where $\rho_{cr} = 3H_0^2/8\pi G \approx 6.8 \times 10^{-30}$ g cm^{-3}, which implies $\rho_\Lambda = \rho_{Pl} \times 10^{-123}$. This shows that the cosmological constant today is 123 orders of magnitude smaller.

This is known as the 'cosmological constant problem'. In the classical big-bang cosmology there is no dynamical theory [8] to relate the cosmological constant to any other physical variable of the universe. There have been some studies [9–11] regarding the universe to relate the space-time manifold to somekind of condensed matter systems. Here by considering [12] the visible universe made up of self-gravitating particles representing luminous baryons and dark matter such as neutrinos (though only a small fraction) which are fermions and a repulsive potential describing the effect of Dark Energy responsible for the accelerated expansion of the universe, we in this chapter derive quantum mechanically a relation connecting temperature, age and cosmological constant of the universe. When the cosmological constant is zero, we get back Gamow's relation with a much better coefficient. Otherwise using as input the current values of $T = 2.728\ K$ and $t = 14 \times 10^9\ years$, we predict the value of cosmological constant as $2.0 \times 10^{-56}\ cm^{-2}$. Note that Λ is a completely free parameter in General Theory of Relativity. Also it is interesting to note that we obtain not only the value of the cosmological constant but also the sign of the parameter correct though it is a very small number.

7.2 Gamow's relation when $\Lambda = 0$

We in this section derive a relation connecting temperature and age of the universe when cosmological constant is zero, by considering a Hamiltonian [12–14] used by us some time back for the study of a system of self-gravitating particles and derived [15] in chapter 2, which is given as:

$$H = -\sum_{i=1}^{N}(\frac{\hbar^2}{2m})\nabla_i^2 + \frac{1}{2}\sum_{i=1}^{N}\sum_{i\neq j,j=1}^{N} v(|\ \vec{X}_i - \vec{X}_j\ |) \qquad (7.1)$$

where $v(|\vec{X}_i - \vec{X}_j|) = -g^2/|\vec{X}_i - \vec{X}_j|$, having $g^2 = Gm^2$, G being the universal gravitational constant and m the mass of the effective constituent particles describing the luminous matter and dark matter whose number is $N = \int \rho(\vec{X})d\vec{X}$. Since the measured value for the temperature of the cosmic microwave background radiation is $\approx 2.728K$, it lies in the neighborhood of almost zero temperature. We, therefore, use the zero temperature formalism for the study of the present problem.

As in the last chapter, We minimize the total energy with respect to λ. Differentiating total energy with respect to λ and then equating it with zero, we obtain the value of λ at which the minimum occurs. This is found as: $\lambda_0 = \frac{72}{25}\frac{\hbar^2}{mg^2}(\frac{3\pi}{16})^{2/3}\frac{1}{N^{1/3}}$ Following the expression for $< KE >$ evaluated at $\lambda = \lambda_0$, we write down the value of the equivalent temperature T of the system, using the relation

$$T = \frac{2}{3k_B}[\frac{< KE >}{N}] = \frac{2}{3k_B}(0.015442)N^{4/3}(\frac{mg^4}{\hbar^2}) \qquad (7.2)$$

The expression for the radius R_0 of the universe, as found by us earlier [14], is given as

$$R_0 = 2\lambda_0 = 4.047528(\frac{\hbar^2}{mg^2})/N^{1/3} \qquad (7.3)$$

Our identification of the radius R_0 with $2\lambda_0$ is based on the use of so called quantum mechanical tunneling [16] effect. Classically, it is well known that a particle has a turning point where the potential energy becomes equal to the total energy. Since the kinetic energy and therefore the velocity are equal to zero at such a point, the classical particle is expected to be turned around or reflected by the potential barrier. From the present theory it is seen that the turning point occurs at a distance $R = 2\lambda_0$.

After invoking Mach's principle [6], which is expressed through the relation $(\frac{GM}{R_0c^2}) \approx 1$, and using the fact that the total mass of the universe $M = Nm$, we are able to obtain the total number of particles N constituting the universe, as

$$N = 2.8535954 (\frac{\hbar c}{Gm^2})^{3/2} \qquad (7.4)$$

Now, substituting Eq.(7.4) in Eq.(7.3), we arrive at the expression for R_0, as

$$R_0 = 2.8535954 (\frac{\hbar}{mc})(\frac{\hbar c}{Gm^2})^{1/2} \qquad (7.5)$$

As one can see from above, R_0 is of a form which involves only the fundamental constants like $\hbar, c, G$ and the effective mass m which is of course not fundamental. Now, eliminating N from Eq.(7.2), by virtue of Eq.(7.4),we have

$$T = \frac{2}{3}(0.0625019)(\frac{mc^2}{k_B}) \qquad (7.6)$$

Since we are considering the visible universe which is actually a patch with a horizon size determined by the speed of light and time that has passed since the bigbang, we now assume that the radius R_0 of the visible universe is approximately given by the relation

$$R_0 \simeq ct \qquad (7.7)$$

where t denotes the age of the universe at any instant of time. Following Eq.(7.5) and Eq.(7.7), we write m as

$$m = (\frac{\hbar^3}{Gc^3})^{1/4}(2.8535954)^{1/2}\frac{1}{\sqrt{t}} \qquad (7.8)$$

It is interesting to see (as shown in Table-1) this variation of mass with time gives approximately the energy and hence the temperature scale of formation of elementary particles in different epochs of nucleosynthesis. We calculate temperatures in different epochs using our Eq.(7.10) to be derived shortly. This is in good agreement with the calculated values of temperature otherwise known from nucleosynthesis calculations [2, 8]. The period between $t = 7 \times 10^{-5}$ sec and 5 sec

TABLE - 1

Age of the universe (t) in sec.	Temperature (T) in K as calculated from Eq.7.10		Temperature (T)in K for the formation of elementary particles [2, 8]
5	$\approx 1 \times 10^9$		$\approx 6 \times 10^9 (e^+, e^-)$
1.2×10^{-4}	$\approx 2.1 \times 10^{11}$		$\approx 1.2 \times 10^{12} (\mu^+, \mu^-$ and their antiparticles)
7×10^{-5}	$\approx 2.8 \times 10^{11}$		$\approx 1.6 \times 10^{12} (\pi^0, \pi^+, \pi^-$ and their antiparticles)
1.5×10^{-6}	$\approx 1.9 \times 10^{12}$		$\approx 10^{13}$ (protons, neutron and their antiparticles)
10^{-43}	$\approx 0.73 \times 10^{31}$		$\approx 10^{32}$ (planck mass)

is called lepton era, while period before $t = 7 \times 10^{-5}$ sec is hadron era and the early era corresponding to the period $t < 10^{-43}$ sec is known as Planck era.

A substitution of m, from Eq.(7.8), in Eq.(7.6), enables us to write

$$T = 0.070388(\frac{1}{k_B})(\frac{c^5\hbar^3}{G})^{1/4}t^{-1/2}$$
$$= 0.06339[\frac{c^2}{G\,a_B}]^{1/4}t^{-1/2} \qquad (7.9)$$

This is exactly the Gamow's relation [8, 12] apart from the fact that Gamow's relation had the coefficient 0.41563 instead of 0.06339 as in our expression. Substituting the numerical value of a_B, which is equal to 7.56×10^{-15} $erg\ cm^{-3}K^{-4}$, and the present value for the universal gravitational constant G [$G = 6.67 \times 10^{-8} dyn.cm^2.gm^{-2}$], in Eq.(7.9),we obtain

$$T = (0.23172 \times 10^{10})t_{sec}^{-1/2}K \qquad (7.10)$$

If we accept the age of the universe to be close to $14 \times 10^9 year$, which we have used here, with the help of Eq.(7.10), we arrive at a value for the Cosmic Microwave Background Temperature (CMBT) equal to $\approx 3.5K$. This is very close to the measured value of 2.728 K as reported from the most recent Cosmic Background Explorer (COBE) satellite measurements [17, 18]. However, if we use Gamow's relation, $t = 956$ billion years is required to obtain the exact value of 2.728 K for the cosmic background temperature from. Using our expression, Eq.(7.10), we would require an age of $22.832 \times 10^9 \ year$ for the universe to get the exact value of 2.728 K. Long back a correction was made to Gamow's relation by multiplying it with a factor of $(\frac{2}{g_d})^{1/4}$ by taking into account the degeneracies of the particles, where $g_d = 9$. This correction effectively multiplies Gamow's relation with a factor of 0.68 and brings back the age of the universe to 425 billion years for the present CMBT. If we multiply our expression by the same factor to correct for the degeneracy of particles, we obtain a value of 2.4 K, which is less than the value of present CMBT. In the next section we see that by including the cosmic repulsion by the part given by cosmological constant we get back 2.728 K, This is physically correct since the cosmological term [6] has the meaning of negative pressure, it adds energy to the system by its tension when the universe expands, though the over all temperature decreases as the universe expands.

7.3 Entropy, number of photons and the ratio $(\bar{N}_\gamma/N_n)$

In this section we estimate total entropy due to the CMBR, total number of photons and the ratio of number of photons to number of baryons. By virtue of the expression given in Eq. (7.9), we can rewrite T as

$$T = 0.070388[(\frac{c^3}{G})(\frac{\pi^2}{60 \ \sigma})]^{1/4}t^{-1/2} \tag{7.11}$$

where $\sigma = \frac{\pi^2 k_B^4}{60\hbar^3 c^2}$ is the Stefan-Boltzmann constant and its numerical value is equal to $5.669 \times 10^{-5} erg/cm^2.K^4.sec$, and we have

$$\sigma T^4 \simeq 2.4547 \times 10^{-5}(\frac{\pi^2 c^3}{60G})\frac{1}{t^2} \tag{7.12}$$

The very form of the above equation suggests that the factor in its right hand side (rhs) can be identified as the energy density of the electromagnetic radiation at a time t. The radiation of this form is believed to follow the black- body law.

Having evaluated the expression in the rhs of Eq. (7.12), the energy of the electromagnetic radiation radiated per unit area per unit time is given as

$$u = 1.6345 \times 10^{33} (\frac{1}{t^2}) \qquad (7.13)$$

where t is the age of the universe in sec at any instant of time. The entropy S associated with the microwave back-ground radiation is obtained as [19]

$$S = \frac{16Vu}{3cT} = 2.9058(\frac{V}{T}) \times 10^{23} (\frac{1}{t^2}) \qquad (7.14)$$

Assuming the present universe to be spherical, its volume V is given as $V = (\frac{4\pi}{3})R_0^3$, where R_0 denotes its radius. Taking $R_0 \simeq 1.325 \times 10^{28}$ cm, which corresponds to the age $t = 14 \times 10^9 yr$, since $(R_0 \approx ct)$, the photonic entropy of the present universe is calculated to give

$$S = 1.45 \times 10^{73} (\frac{1}{T}) erg/deg \qquad (7.15)$$

For $T = 2.728K$, it becomes,

$$S = 0.5 \times 10^{73} \approx 10^{73} erg/deg \qquad (7.16)$$

The equilibrium number of photons [19] associated with the microwave background radiation is given as

$$\overline{N}_\gamma = \frac{V2\zeta(3)}{\pi^2 \hbar^3 c^3} k_\beta^3 T_0^3 \simeq (410.0)V \qquad (7.17)$$

Following this, the photon density is found to be $(\frac{\overline{N}_\gamma}{V}) \simeq 410$, which is in very good agreement with the estimated value of 400 found [20] by doing a calculation of the total energy density carried by the cosmic microwave background radiation. Using Eq. (7.17), we have calculated the total number of photons in the present universe, which becomes

$$\overline{N}_\gamma = 0.4 \times 10^{88} \qquad (7.18)$$

175

Considering the fact that the number of nucleons [21], N_n, in the present universe is $\approx 6.30 \times 10^{78}$, we obtain

$$(\frac{\overline{N_\gamma}}{N_n}) \simeq 0.063 \times 10^{10} \tag{7.19}$$

This agrees with the value $(0.14 \sim 0.33) \times 10^{10}$ as speculated by several earlier workers [22] following calculations on baryogenesis and our calculated value in chapter 6. So we find that our theory reproduces the temperature of the cosmic background radiation correctly. Besides, it also succeeds to reproduce the photon density associated with the background radiation, and the value of the ratio $(\bar{N}_\gamma/N_n)$, which nicely match with the results predicted by others.

7.4   A relation connecting t, T and Λ

The cosmological constant term [5, 6] Λ associated with vacuum energy density was originally introduced by Einstein as a repulsive component in his field equation and when translated from the relativistic to Newtonian picture gives rise to a repulsive harmonic oscillator force per unit mass as $\sim (\Lambda c^2)\vec{r}$ between points in space when Λ is positive. The one-body operator corresponding to the potential can be written as $H_\Lambda = -\Lambda c^2 |\vec{X}|^2 \rho(\vec{X})$ where $\rho(X)$ here is measured in the unit of mass density and this term also contains a unit volume. Hence the energy corresponding to this repulsive potential can be written as:

$$< H_\Lambda >= - \int \Lambda c^2 |\vec{X}|^2 \rho^2(\vec{X}) \, d\vec{X} \tag{7.20}$$

By including this contribution of H_Λ in the total energy expression calculated earlier, we have the new total energy

$$E(\lambda) = \frac{\hbar^2}{m} \frac{12}{25\pi} (\frac{3\pi N}{16})^{5/3} \frac{1}{\lambda^2} \quad \frac{g_\Lambda^2 N^2}{16} \cdot \frac{1}{\lambda} \tag{7.21}$$

where $g_\Lambda^2 = g^2 + \frac{3\Lambda c^2}{16\pi}$ and dimension of first term is same as the second term since the second term contains implicitly a product of unit mass and unit volume as can be easily seen by checking the single particle Hamiltonian H_Λ. Calculating as before, we have

$$N = 2.8535954(\frac{1}{Gm^2})^{3/4}(\frac{\hbar c}{g_\Lambda})^{3/2} \qquad (7.22)$$

and

$$R_0 = 2.8535954(\frac{\hbar}{mc})^{1/2}(\frac{\hbar G^{1/4}}{g_\Lambda^{3/2}}) \qquad (7.23)$$

Now equating this R_0 with ct we have

$$Gm^{8/3} + \frac{3\Lambda c^2}{16\pi}m^{2/3} - Q = 0 \qquad (7.24)$$

where $Q = \frac{4.0475279\hbar^2 G^{1/3}}{c^2} \times \frac{1}{t^{4/3}}$. Using $m' = m^{2/3}$, the above equation can be cast as a quartic equation in m'. We find [23] four analytic solutions for m' and hence for m. Three of the solutions are unphysical and the only solution which is physically correct is given as

$$m = (\frac{u^{1/2} + \sqrt{u - 4(u/2 - [(u/2)^2 + Q/G]^{1/2})}}{2})^{3/2} \qquad (7.25)$$

where

$$u = [r + (q^3 + r^2)^{1/2}]^{1/3} + [r - (q^3 + r^2)^{1/2}]^{1/3} \qquad (7.26)$$

and $r = \frac{9\Lambda^2 c^4}{2(16\pi G)^2}$, $q = \frac{4Q}{3G}$. Now the Kinetic energy with the degeneracy factor as discussed in the previous section, is given as

$$T = (\frac{2}{g_d})^{\frac{1}{4}}\frac{2}{3k_B}[\frac{<KE>}{N}] = (\frac{2}{g_d})^{\frac{1}{4}}\frac{2}{3k_B}(0.015442)N^{4/3}(\frac{mg_\Lambda^4}{\hbar^2}) \qquad (7.27)$$

Using Eq.(7.22) and Eq.(7.25) in Eq.(7.27), we finally have the relation,

$$T = 0.0417(\frac{2}{g_d})^{1/4}\frac{c^2}{k_B}\frac{[(\{u^{1/2} + \sqrt{4[(u/2)^2 + Q/G]^{1/2} - u}\}/2)^3 + \frac{3\Lambda c^2}{16\pi G}]}{(\{u^{1/2} + \sqrt{4[(u/2)^2 + Q/G]^{1/2} - u}\}/2)^{3/2}} \qquad (7.28)$$

This is the central result of our paper. This relation connects temperature T with time t and cosmological constant Λ since Q is a function of t and u is also a function of t and Λ. When $\Lambda=0$, we get back the relation Eq.(7.9) connecting T and t. Since we know the current values of $T = 2.728K$ and $t = 14 \times 10^9 year$, using Eq.(7.28), we solve for Λ. We do that in Fig.7.1 by plotting the left hand side and right hand side of Eq. (7.28) and finding the crossing point. This gives $\Lambda = 2.0 \times 10^{-56}$ cm^{-2} which is the value that has been derived dynamically here.

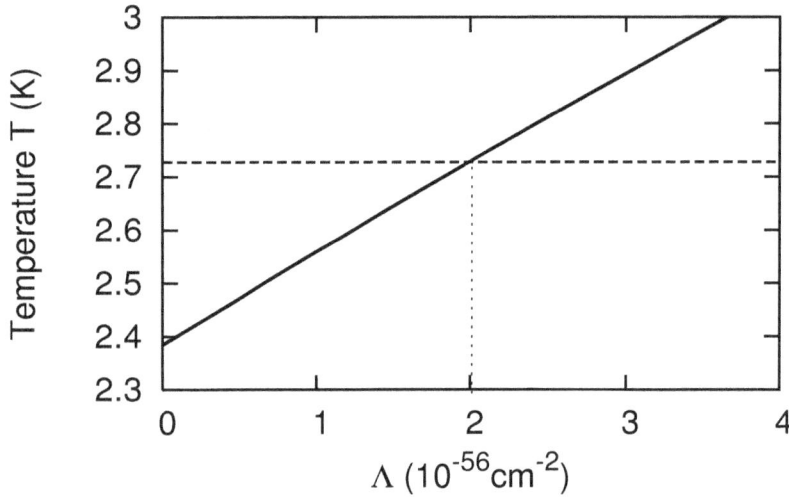

Figure 7.1: Determination of Λ by plotting the right hand side of Eq.(7.28) as a function of Λ (solid line) and left hand side as 2.728 K (thin broken line). The vertical dotted line indicates the value of $\Lambda = 2.0 \ 10^{-56} cm^{-2}$

7.5 Summary

Considering our expanding universe as made up of gravitationally interacting particles which describe particles of luminous matter, dark matter and dark energy which is described by a repulsive harmonic potential among the points in the flat 3-space, we derive a quantum mechanical relation connecting, temperature of the cosmic microwave background radiation, age, and cosmological constant of the universe. When the cosmological constant is zero, we get back Gamow's relation with a much better coefficient. Otherwise, our theory predicts a value of the cosmological constant 2.0×10^{-56} cm^{-2} when the present values of cosmic microwave background temperature of 2.728 K and age of the universe 14 billion years are taken as input. It is interesting to note that in this flat universe, our method dynamically determines the value of the cosmological constant reasonably well compared to General Theory of Relativity where the cosmological constant is a free parameter.

Bibliography

[1] A. H. Guth, in *Bubbles, voids and bumps in time: the new cosmology* edited by J. Cornell (Cambridge University Press, Cambridge, 1989).

[2] G. Contopoulos and D. Kotsakis, *Cosmology* (Springer, Heidelberg, 1987).

[3] A. G. Riess *et al.* [Supernova Search Team Collaboration], Astron. J. **116** 1009 (1998).

[4] S. Perlmutter *et al.* [Supernova Cosmology Project Collaboration], Astrophys. J. **517** 565 (1999).

[5] P. J. E. Peebles and B. Ratra, Rev. Mod. Phys. **75**, 559 (2003).

[6] E. Harrison, *Cosmology* (Cambridge University Press, Cambridge, 2000).

[7] S. M. Carroll, Living Reviews in Relativity, **4**, 1 (2001).

[8] J. V. Narlikar, *An introduction to cosmology, 3rd Ed.* (Cambridge University Press, Cambridge, 2002).

[9] G. E. Volovik, *The Universe in a Helium Droplet* (Clarendon, Oxford, 2003).

[10] R. B. Laughlin, Int. J. Mod. Phys. A, **18**, 831 (2003).

[11] J. P. Hu and S. C. Zhang, Phys. Rev. B. **66**, 125301 (2002).

[12] S. Mishra, Int. J. Theor. Phys. **47**, 2655 (2008).

[13] D. N. Tripathy and S. Mishra, Int. J. Mod. Phys. D **7**, 3 , 431 (1998).

[14] D. N. Tripathy and S. Mishra, Int. J. Mod. Phys. D **7**, 6, 917 (1998).

[15] J. Christian, Phys. Rev. D, **56**, 4844 (1997).

[16] A. Karplus and R. N. Porter, *Atoms and Molecules* (Benjamin, Reading MA 1970).

[17] A. A. Penzias and R. W. Wilson, *Astrophys. Jour.* **142**, 419 (1965).

[18] D. J. Fixen, E. S. Cheng, J. M. Gales, J. C. Mather, R. A. Shafer and E. L. Wright, Astrophysics. J. **473**, 576 (1996); A. R. Liddle, Contemporary Physics, **39**, no 2, 95 (1998).

[19] R. K. Pathria, *Statistical Mechanics* (Pergamon Press, Oxford, 1972).

[20] A. M. Boesgaard and G. Steigman, Ann. Rev. Astron. **23**, 319 (1985).

[21] P. S. Wesson, *Cosmology and Geophysics* (Adam Hilger, Bristol, 1978).

[22] I. Affleck and M. Dine, Nucl. Phys. **B249**, 361 (1985).

[23] M. Abramowitz and I. A. Stegun, *Handbook of Mathematical Functions* (Dover, New York, 1965).

Chapter 8

Summary

In this book we established an Effective Theory of Quantum Gravity. First it was shown that the Newton-Cartan theory which is the non-special relativistic limit of GTR, with vanishing spatial curvature can be quantized exactly and it gives rise to a Hamiltonian which is well known otherwise. We special relativized the problem and studied the self-gravitating systems using that quantum many-particle Hamiltonian. The systems are stars like neutron star, white dwarf and black hole. Also the universe was treated with the same approach.

Though the Hamiltonian used in the analysis can be naively thought of as a truncated (without higher order self-interaction terms) and hence an approximate one at the first sight, actually it is now known as the exact Hamiltonian of a soluble sector of quantum gravity where the spacial curvature is zero. This represents the present universe very appropriately where the above curvature is known to be zero. In case of other self-gravitating systems discussed in the book the Hamiltonian is an approximate one since the spacial curvature is finite for these cases.

We use variational approach using an trial density to calculate the energy of each of these systems. We relativize the problem to include speed of light through the Schwarzschild radius in case of different stars and through the Mach's principle in case of the universe.

In case of white dwarfs, we obtain the Chandrasekhar limit and also in case of neutron star we get a maximum mass beyond which the gravitational collapse is complete and the neutron star becomes a black hole. We show that other parameters of stars like density, temperature, size etc that we calculate from our theory are reasonably very good. Incase of black hole we derive the Schwarzschild radius quantum mechanically and including an higher order term in the self-energy we find a correction to this radius.

Our approach produces most of the parameters of the universe such as number of baryons, total mass density and an upper limit on the mass of neutrino. We reproduce the Gamow's relation between time and temperature of cosmic back ground radiation of the universe. When the repulsive force responsible for the accelerated expansion of the universe is included through the Newtonian limit of the Einstein's cosmological constant term we get a relation connecting time, temperature of the cosmic back ground radiation and the cosmological constant. Knowing the present value of the age and temperature of the CMBT we solve for the cosmological constant and find a value which is very good as compared with the present observational data.

Appendix A

Total energy of a many-particle system

The Hamiltonian of a classical system of N self-gravitating particles each having a mass m and momentum $\vec{p}_i$ is given as

$$H = \sum_{i=1}^{N} \frac{\vec{p}_i^2}{2m} - \sum_{i>j}^{N} \frac{Gm^2}{|\vec{r}_i - \vec{r}_j|} \tag{A.1}$$

The Schrodinger equation for this N-particle system with wave function $\Psi(1, ..., N)$ and $\vec{p} = -i\hbar\vec{\nabla}$ can be written as

$$H\Psi(1, ..., N) = E\Psi(1, ..., N). \tag{A.2}$$

If we consider the N-particle system as a system of fermions, then the total wavefunction is given as a Slater determinant as

$$\Psi(1, ..., N) = \frac{1}{\sqrt{N!}} \sum_{p} (-1)^P P\psi_1(1)....\psi_N(N) \tag{A.3}$$

where the single particle wavefunction which is a product of spacial and spin states is $\psi_i(i) = \psi_i(\vec{r}_i)\chi_i(m_{s_i})$ and it is normalized as $\int d^3\vec{r}|\psi_i(\vec{r})|^2 = 1$.

Now taking the expectation value of the total Hamiltonian H over the state Ψ, we have the total energy

$$E = <H> = \sum_{i}^{N} \int d^3\vec{r} \left[-\frac{\hbar^2}{2m}|\nabla \psi_i(\vec{r})|^2 \right]$$

$$-\frac{1}{2}\sum_{i,j,i\neq j}^{N}\int d^3\vec{r}\int d^3\vec{r'}\frac{Gm^2}{|\vec{r}-\vec{r'}|}|\psi_i(\vec{r})|^2|\psi_i(\vec{r'})|^2$$

$$+\frac{1}{2}\sum_{i,j,i\neq j}^{N}\delta_{m_{s_i}m_{s_j}}\int d^3\vec{r}\int d^3\vec{r'}\frac{Gm^2}{|\vec{r}-\vec{r'}|}\psi_i^*(\vec{r})\psi_i(\vec{r'})\psi_j^*(\vec{r'})\psi_j(\vec{r}), \qquad (A.4)$$

where the first term is the total kinetic energy, second term is the direct interaction energy and the third term is known as the exchange interaction energy. The above energy can be written [1] in terms of density $\rho(r)$, such that $\int d^3\vec{r}\,\rho(r) = N$, as:

$$E = E_k + E_d + E_{ex}$$

$$= \int d^3\vec{r}\left[\frac{3}{5}\frac{\hbar^2\pi^2}{2m}\left(\frac{3}{\pi}\right)^{2/3}\rho^{5/3}(\vec{r})\right]$$

$$-\frac{1}{2}\int d^3\vec{r}\int d^3\vec{r'}\frac{Gm^2}{|\vec{r}-\vec{r'}|}\rho(\vec{r})\rho(\vec{r'})$$

$$+\int d^3\vec{r}\frac{3}{4}Gm^2\left(\frac{3}{\pi}\right)^{1/3}\rho(r)^{4/3} \qquad (A.5)$$

Appendix B

Scale factor

In this appendix we derive the evolution of the scale factor according to GTR. The Maximally symmetric Robertson-Walker metric for a homogeneous and isotropic universe [2] is given as

$$ds^2 = dt^2 - a^2(t)\{\frac{dr^2}{1 - kr^2} + r^2 d\theta^2 + r^2 sin^2\theta d\phi^2\} \tag{B.1}$$

where $k = 1/R^2$ is the curvature constant and a(t) is the scale factor. The Einstein equation is given as

$$R_{\mu\nu} - \frac{1}{2}g_{\mu\nu}R + \Lambda g_{\mu\nu} = -8\pi G T_{\mu\nu} \tag{B.2}$$

where $R_{\mu\nu}$ is the Ricci tensor, R is Ricci scalar, $g_{\mu\nu}$ the metric, Λ the cosmological constant. $T_{\mu\nu}$,the energy-momentum tensor is taken as

$$T_{\mu\nu} = diag[\rho, -p, -p, -p] \tag{B.3}$$

One can get from Einstein equation the following two equations which govern the evolution of the scale factor $a(t)$ as

$$\frac{\dot{a}^2}{a^2} + \frac{k}{a^2} = \frac{8\pi G}{3}(\rho + \rho_\Lambda) \tag{B.4}$$

$$\frac{2\ddot{a}}{a} + \frac{\dot{a}^2}{a^2} + \frac{k}{a^2} = \frac{8\pi G}{3}(-p + \rho_\Lambda) \tag{B.5}$$

where $\rho_\Lambda = \Lambda/(8\pi G)$. These two equations are known as Einstein-Friedmann-Lemaitre equations [3, 4]. Since the Hubble constant is defined as $H = v(t)/r(t) = \dot{a}/a$, Eq. (B.4) can be written in terms of the Hubble constant as,

$$\frac{k}{H^2 a^2} = \frac{8\pi G}{3H^2}(\rho + \rho_\Lambda) - 1 = \Omega - 1 \tag{B.6}$$

where $\Omega = \Omega_m + \Omega_\Lambda$ and $\Omega_m = \rho/\rho_c$, $\Omega_\Lambda = \rho_\Lambda/\rho_c$ and the critical density $\rho_c = 3H^2/8\pi G$.

We see that for $k = +1, 0, -1$ we have $\Omega >, =, < 1$ respectively. The two Einstein-Friedmann-Lemaitre equations when combined give

$$\frac{\ddot{a}}{a} = -\frac{4\pi G}{3}(\rho + 3p) + \frac{\Lambda}{3} \tag{B.7}$$

The first figure in (B.1) describes the evolution of our universe which has zero curvature and is expanding with acceleration due to a finite cosmological constant Λ.

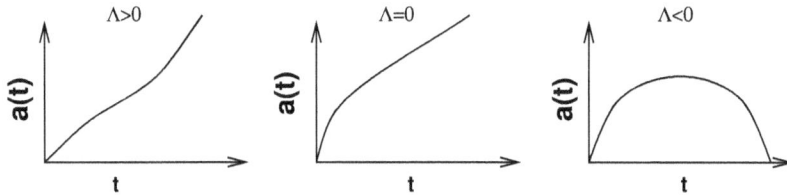

Figure B.1: The scale factor for zero curvature $k = 0$ and different values of cosmological constant.

Appendix C

The principle of extremal action

The principle of extremal action is a variational principle which gives the equation of motion by varying the action of the theory. The action is a functional of the Lagrangian of the system. The Lagrangian is defined as the difference of kinetic energy and potential energy.

$$\mathcal{L}(\dot{\phi}, \phi) = \mathcal{KE}(\dot{\phi}) - \mathcal{PE}(\phi) \tag{C.1}$$

The action is defined as

$$\mathcal{S}[\phi(t)] = \int_{t_a}^{t_b} dt \mathcal{L}(\dot{\phi}(t), \phi(t)) \tag{C.2}$$

The path that extremize the action gives the equation of motion. The vanishing of the first-order variation $\delta\mathcal{S}[\phi(t)]$ for any arbitrary variation $\delta(t)$ of the path (ϕ_a, t_a) to (ϕ_b, t_b) leads to the extrema of the functional $\mathcal{S}[\phi(t)]$. In order to get $\delta\mathcal{S}[\phi(t)]$, we substitute $\phi(t)$ with $\phi(t) + \delta\phi(t)$ in $\mathcal{S}$ and expand to first order in $\delta\phi(t)$ and integrate to get

$$
\begin{aligned}
\delta\mathcal{S}[\phi(t)] &= \int_{t_a}^{t_b} dt \left[\frac{\partial\mathcal{L}}{\partial\dot{\phi}(t)} \delta\dot{\phi}(t) + \frac{\partial\mathcal{L}}{\partial\phi(t)} \delta\phi(t) \right] \\
&= \frac{\partial\mathcal{L}}{\partial\dot{\phi}(t)} \delta\phi(t)\big|_{t_a}^{t_b} + \int_{t_a}^{t_b} dt \left[-\frac{d}{dt}\left(\frac{\partial\mathcal{L}}{\partial\dot{\phi}(t)}\right) + \frac{\partial\mathcal{L}}{\partial\phi(t)} \right] \delta\phi(t). \quad \text{(C.3)}
\end{aligned}
$$

The first term vanishes trivially and vanishing of the second for arbitrary $\delta\phi(t)$ gives us that the integrand should vanish.

$$-\frac{d}{dt}\left(\frac{\partial\mathcal{L}}{\partial\dot{\phi}(t)}\right) + \frac{\partial\mathcal{L}}{\partial\phi(t)} = 0 \tag{C.4}$$

As an example consider the Lagrangian

$$\mathcal{L}(\dot{\phi},\phi) = \frac{1}{2}m\dot{\phi}^2 - V(\phi) \tag{C.5}$$

then using the Eq.(C.4), we get the equation of motion as Newton's law $m\ddot{\phi} = -dV/d\phi$.

Appendix D

Some constants of nature

Speed of light	$c = 2.99792 \times 10^{10} \; cm \; s^{-1}$
Gravitational constant	$G = 6.6720 \times 10^{-8} \; dyn \; cm^2 \; g^{-2}$
Planck constant	$\hbar = 1.0545 \times 10^{-27} \; erg \; s$
Boltzmann constant	$k = 1.38066 \times 10^{-16} \; erg \; K^{-1}$
Electron mass	$m_e = 0.5110 \; MeV$
Proton mass	$m_n = 938.2796 \; MeV$
Planck length	$\sqrt{(G\hbar/c^3)} = 1.62 \times 10^{-33} \; cm$
Planck time	$\sqrt{(G\hbar/c^5)} = 5.39 \times 10^{-44} \; s$
Planck mass	$\sqrt{(c\hbar/G)} = 2.2 \times 10^{-5} \; g$
Planck energy	$\sqrt{(c^5\hbar/G)} = 1.22 \times 10^{19} \; GeV$
Planck density	$\sqrt{(c^5/\hbar G^2)} = 5.16 \times 10^{93} \; g \; cm^{-3}$
Lightyear	$1 \; ly = 9.4605 \times 10^{17} \; cm$
Parsec	$1 \; pc = 3.26 \; ly$
Mass of the Sun	$M_\odot = 1.989 \times 10^{33} \; g$
Hubble constant	$H_0 = 70 \; km \; s^{-1} \; Mpc^{-1}$
Age of the universe	$T_0 = 14 \times 10^9 \; yr$

Bibliography

[1] H. A. Bethe and R. W. Jackiw, *Intermediate Quantum Mechanics* (Benjamin, London, 1968).

[2] J. B. Hartle, *Gravity* (Pearson Education, 2003).

[3] A. Friedmann, Z. Phys. **10**, 377 (1992); Z. Phys. **21**, 326 (1924).

[4] A.J. Lemaitre, *Mon. Not. Roy. Astron. Soc.* **91** (1931).

[5] W.-M. Yao et al (Particle data group), J. Phys. **G 33**, 1 (2006).

www.ingramcontent.com/pod-product-compliance
Lightning Source LLC
Chambersburg PA
CBHW081524220326
41598CB00036B/6324